Energy, the Environment, and Public Opinion

Energy, the Environment, and Public Opinion

Eric R. A. N. Smith

ROWMAN & LITTLEFIELD PUBLISHERS, INC.
Lanham • Boulder • New York • Oxford

ROWMAN & LITTLEFIELD PUBLISHERS, INC.

Published in the United States of America
by Rowman & Littlefield Publishers, Inc.
4720 Boston Way, Lanham, Maryland 20706
www.rowmanlittlefield.com

12 Hid's Copse Road, Cumnor Hill, Oxford OX2 9JJ, England

British Library Cataloguing in Publication Information Available

Library of Congress Cataloging-in-Publication Data

Smith Eric R. A. N.
    Energy, the environment, and public opinion / Eric R. A. N. Smith.
      p. cm.
    Includes Bibliographical references and index.
    ISBN 0-7425-1025-5 (alk. paper)—ISBN 0-7425-1026-3 (pbk. : alk. paper)
    1. Energy development—Environmental aspects—United States—Public opinion. 2. Energy policy—United States—Public opinion. 3. Public opinion—United States. I. Title.

TD195.E49 S66 2001
333.79'14'0973—dc21   2001034619

Printed in the United States of America

♾ ™ The paper used in this publication meets the minimum requirements of American National Standard for Information Sciences—Permanence of Paper for Printed Library Materials, ANSI/NISO Z39.48-1992.

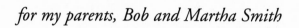

for my parents, Bob and Martha Smith

Contents

Figures and Tables

FIGURES

TABLES

Acknowledgments

Like all books, this book benefited from the contributions of many people. I would especially like to thank the graduate students who have worked and coauthored papers with me on environmental politics—Sonia Garcia, Marisela Marquez, Nina Van Dyke, Bill Herms, and Juliet Carlisle. They are all outstanding scholars, and I am grateful for their help. Many of their contributions can be seen in these pages.

Other friends read portions of the manuscript or discussed the ideas with me. They gave me many useful suggestions and helped make this a better book than it would have been without their assistance. For their help and encouragement, I would like to thank Steve Weatherford, Tim Alison, Robert O'Connor, Bill Freudenburg, Fred Thompson, Michael Delli Carpini, Jack Citrin, Harvey Molotch, John Woolley, and Jim Lima. I would also like to thank Biliana Cicin-Sain and Dean Mann for urging me to turn toward research on environmental politics.

This work would not have been possible without the generous financial support from the Minerals Management Service, U.S. Department of the Interior, under MMS Agreements No. 14-35-0001-03471, 14-35-0001-30471, and 14-32-000130761. MMS would like me to add that the views and conclusions contained in this document are those of the author and should not be interpreted as necessarily representing the official policies, either express or implied, of the U.S. government. Official disclaimers aside, I would like to thank my friends at MMS—Jim Lima, Fred Piltz, Fred White, and Rodney Cluck for their friendship and support.

This book would also not have been possible without the California Offshore Oil Drilling and Energy Policy Survey, which was funded by the University of California Toxic Substances Education and Research Program. I received that grant in large part because of the efforts of Harvey Molotch, who deserves special thanks.

I would also like to thank Mark DiCamillo for his help in designing the California Offshore Oil Drilling and Energy Policy Survey. Of course, neither he nor the Field Institute bears any responsibility for any mistakes in design or interpretation. More generally, Mark DiCamillo, Mervin Field, and the Field Institute staff deserve thanks for their generous assistance to academics, politicians, policymakers, and journalists since the 1950s. Without their Field Polls, neither this book nor many other pieces of research would have been written.

For their help with numerous grant applications, I would also like to thank the leaders and staff of U.C. Santa Barbara's Marine Science Institute—Russ Schmitt, Bonnie Williamson, Shanna Bowers, Marie Ciluaga, and Judy McCaslin. And for their assistance with computer and data problems throughout this project, I would like to thank Joan Murdoch, Lia Roberts, and Steve Velasco. The editors at Rowman & Littlefield, Jennifer Knerr and Janice Braunstein, provided excellent editorial advice.

Finally, I would like to thank my wife, Elizabeth, and our daughters, Katie and Steffi, for their love and support. And I would like to thank my parents, Bob and Martha Smith, for their love and for teaching me to enjoy the wilderness. This book is dedicated to them.

1

Introduction

On 26 March 1996, the voters of Santa Barbara County, California, approved Measure A—a ballot initiative requiring a public vote to allow any new oil development project in the county. For all practical purposes, the passage of that initiative ended any hope for new oil development in Santa Barbara. The oil companies recognized that it would be nearly impossible to win a public vote of approval for a new oil development. The Mobil Oil Corporation, which had been seeking county permission for a new, extended-reach drilling operation to tap into an offshore oil pool, gave up its effort, closed down its main Santa Barbara office, and left town.

That same winter of 1996, gasoline prices across the United States began rising sharply for the first time since the Persian Gulf War in 1990. In southern California, the cost of a gallon of gas rose from $1.15 in December to $1.47 in April, with similar price hikes occurring all across the nation.[1] The jump in prices bought a quick response from the public—it was not happy. Because 1996 was a presidential election year, the jump also brought a quick response from politicians. Republican presidential candidate and Senate Majority Leader Bob Dole called for a repeal of the 4.3-cents-per-gallon gas tax that President Bill Clinton had proposed and Congress had passed in the early months of 1993. President Clinton, himself a candidate for reelection, ordered the release of millions of barrels of oil from the U.S. strategic oil reserve in order to increase the supply of oil going to the nation's refineries and to cut gasoline prices. In a burst of election-year speed, the House passed the gas-tax cut in May, and the Senate followed in June, with the president signing it shortly afterward.[2]

These two events—the Santa Barbara vote ending local oil development, and the public demand for government action to cut gasoline prices—illustrate a critical aspect of the politics of the energy situation facing the United States. On the one hand, the public generally opposes new energy development. No energy source is popular. Nuclear power may be the least-liked form of energy production, but oil, coal, and hydroelectric power are hardly popular either. On the other hand, the public does not want to go without cheap, plentiful energy, nor does it fondly view the prospect of taking further action to conserve the nonrenewable energy sources we have. In short, the public wants to consume its energy cake and keep it undeveloped at the same time.

If the historically low cost of oil in the world markets during the late 1990s had continued indefinitely, Americans' attitudes toward energy problems might have made sense. We could have preserved our stunning coastlines by banning offshore oil drilling, avoided worrying about nuclear meltdowns by preventing the construction of new nuclear power plants, and imported all the oil we wanted at ridiculously low prices. The problem is that the situation did not last—and in fact, could not last.

In the early days of 2000, oil prices once again rose sharply. By late summer, large numbers of Californians had to deal with the threat of rolling blackouts because of insufficient power in the western states' electrical power grid. Of course, this energy crisis might only be temporary, lasting no more than a year or two. Yet by some time in the first half of this century, our energy problems will no longer be temporary. We will begin to face continuing energy problems that will not go away.

In this century, America will have to come to grips with two facts—that using petroleum and other fossil fuels contributes to global warming, and that the world's petroleum supply will probably run out by the end of the century. How the public responds to these facts, as they become more obvious, will have a huge impact on U.S. energy policy and on the American economy.

This is a book about public opinion toward energy issues. In it, I investigate what the public thinks about oil, nuclear power, and other energy options. I address such questions as, how have people's opinions changed over time, and how are they likely to change in the future? How much does the public understand about energy issues? Do people know enough to assess the facts rationally and to reason about various policy options? Who favors further oil development or the expansion of nuclear power? How do opinions differ from group to group? Does it matter how close one lives to an oil-drilling operation or nuclear power plant? What roles do ideology and other beliefs play in influencing opinions on energy issues? Taken together, the answers to these questions should help us understand both public opinion on energy issues and how people are likely to react to the energy problems of the coming century.

Understanding public opinion about energy issues is critical—because of the importance of energy to our economy and our environment, and because of the important role public opinion plays in setting energy policy. If the public demands that Congress enact foolish policies, we may all be in trouble. Let me briefly expand on these points.

ENERGY POLICY AND THE ENVIRONMENT

The American economy is built on cheap, abundant energy. Throughout the twentieth century, the low-cost and seemingly endless supply of energy helped transform America from an agricultural society to an industrial one and then to a postindustrial one. Although cheap energy was not the only cause of the rapid expansion of the American economy, it was certainly one of the critical factors.[3] By the 1970s, Americans consumed more energy per capita than citizens of any other nation—leading historian David Nye to describe our economy as the "high-energy economy."[4]

Energy is not only enormously important to America's economy, it is of central importance to a host of environmental issues as well. The 1969 blowout of a Union Oil Company offshore oil-drilling platform in the Santa Barbara Channel helped launch the modern environmental movement. The blowout dumped hundreds of thousands, perhaps millions, of gallons of oil into the channel, creating an oil slick that covered eight hundred square miles of ocean and shoreline and grabbing headlines around the world.[5] The nightly television footage of thousands of dead fish and oil-coated birds sparked a massive public reaction and helped pressure President Richard Nixon and Congress to pass a wide range of environmental legislation.[6] Thousands of people joined environmental groups, and millions more began to realize the possible dangers of oil production and of insufficient regulation of industries with a potential for environmental threats.

If the Santa Barbara oil spill launched the environmental movement, the energy crisis kept it going. Robert Paehlke writes, "The historic event most central to environmentalism was the energy price shock of 1973 and 1979."[7] Few would dispute him. The massive price hikes and the long lines at the gas stations forced energy issues to the forefront of American politics. Congress and the White House responded with a series of programs, from gas rationing to downsizing cars to increasing energy production. Many of these programs sparked controversy, either because they limited people's freedom to do what they wished with their property or because they damaged environment. The result was to focus the public's attention on a wide range of environmental disputes.

Looking back at the years since the first energy crisis, we can see a long list of bitterly contested issues stemming from our energy problems. In the last thirty years, Americans have fought over whether to impose fuel-efficiency standards on automobiles; whether to build nuclear power plants, whether to drill for oil along their coastlines and in other environmentally sensitive areas, such as the Arctic National Wildlife Reserve; whether to build an oil pipeline across Alaska; whether to allow strip mining for coal in western states; whether to build power dams on some of America's last wild rivers; whether to allow the largest coal-fired power plant in the United States to continue to pollute the air of the Grand Canyon; and where to locate toxic and nuclear waste dumps. The list goes on and on.

The battles over energy policy will surely continue into the next century. Two critical areas of conflict stand out—what to do about global warming, and what to do as the world supply of petroleum begins to run out and prices begin to rise. Between these two problems, global warming has received the lion's share of attention to date. Ever since the unusually hot, dry summer of 1988, global warming has been the subject of growing concern among scientists, lawmakers, and the public.[8]

Earlier in the century, global warming was no more than a worrisome but unproven hypothesis. Scientists had been studying the earth's climate since the late 1800s, and they realized that human activity could cause the earth's average temperature to rise—at least in principle. They also realized that the earth's temperature could fluctuate for other reasons. The existence of the ice ages, after all, was well established before the turn of the century. Then in 1988, just as scientific understanding of the climate was beginning to solidify, the United States had a summer packed with unusually severe droughts and record temperatures. The results were public attention to the issue and sharply increased scientific funding to investigate it. The Intergovernmental Panel on Climate Change was formed that year, and the stage was set for the 1992 United Nations Conference on Environment and Development—the "Earth Summit," as it was popularly called. Now, as the new millennium opens, the evidence on global warming is mounting, and atmospheric scientists have come to a consensus that carbon dioxide and other greenhouse gases are, in fact, warming the earth.[9]

Much of the debate about global warming will center on energy issues because the burning of fossil fuels produces the bulk of the greenhouse gases that allegedly contribute to global warming. There are no easy solutions. The expansion of nuclear power would help diminish world reliance on fossil fuels and lessen global warming, but most environmentalists would vehemently resist it. Other alternatives, such as developing various renewable energy sources and following the "soft energy path" described by Amory Lovins, would entail substantial changes in the way our economy and society are organized, and they, too, will be bitterly resisted.[10]

To complicate matters, the problem of global warming is intimately related to the problem of the diminishing world supply of petroleum. The world is, in fact, running out of oil—a claim I shall explain and defend in the next chapter—and this will have profound consequences. In the coming century, energy prices will rise and the world will have to convert to substitutes for oil. The two most obvious substitutes—coal and nuclear power—can seriously harm the environment. Burning coal produces both air pollution and carbon dioxide, the principal greenhouse gas. Nuclear power comes with the immediate possibility of deadly radiation leaks and the long-term problem of disposing of the used fuel, much of which has a half-life of a mind-boggling twenty-four thousand years.[11]

Whether the world can develop soft-path alternatives without major economic consequences remains to be seen. Despite fervent hopes by environmentalists and a good deal of government regulation, the American public has never been committed to the idea of conservation. Recycling aluminum cans is fine, but passing up on an opportunity to buy a gas-guzzling SUV is not. Per capita consumption of energy fell during the energy crises of the 1970s, but since then it has risen to record levels.

There have also been many brave predictions about a wide range of renewable sources of energy, such as wind and solar power, but so far those claims have not proven true. Most predictions about new energy sources have turned out to be wildly exaggerated or just plain wrong. All the types of alternative energy together still account for only about one half of 1 percent of American energy consumption. Of course, environmentalists have not been the only ones to see all our problems as easily solved in the future. In 1954 Lewis Strauss, the chair of the Atomic Energy Commission, confidently proclaimed, "Our children will enjoy energy too cheap to meter" as a result of nuclear power.[12] The only thing we can safely predict is that public policy fights over energy policy will continue for many decades to come.

THE ROLE OF PUBLIC OPINION

What the public thinks about energy issues is important because public opinion is a major force in politics. Some skeptics think that despite the fact that we live in a democracy, the voice of the people is rarely influential. To them, elections are shows put on for symbolic reasons, with little effect on public policy.[13] Others believe that in technically complex areas such as energy policy, the public ought not to have any impact. To these critics, such technical questions as the safety of nuclear power plants should be left to technical experts. If the nuclear power industry's experts declare them safe, that should be the end of it.[14] Yet the skeptics' beliefs are false and the critics' wishes will never come true.

Public opinion has a major impact on energy policy because politicians generally do what their constituents want on major issues. The fact that elected officials generally follow their constituents' preferences is well documented.[15] Academic studies aside, one can easily see the influence of public opinion in the rapid demise of the nuclear power industry during the 1970s and in the moratorium on offshore oil drilling along many coastal areas around the nation that began in 1990.[16] The public did not like either energy source, and despite the power and lobbying influence of the nuclear power and oil industries, the public got what it wanted. Public opinion was not the only cause of the problems faced by these industries, especially in the case of nuclear power, but it was a powerful contributing cause of the end of the expansion of both nuclear power and coastal oil development. One can also see the influence of public opinion in every energy crisis from the first one in 1973–1974 to our current crisis. Every time the public demanded lower energy prices, politicians jumped to do the public's bidding—despite the fact that lower energy prices encouraged consumption and actually made the energy shortages worse.[17]

The initiative process allows the public more direct control over public policy. In the twenty-three states and hundreds of cities and counties with initiative process, the public can not only express opinions, it can pass laws. In California, for example, voters have passed both local and statewide ballot initiatives limiting onshore and offshore oil development. Although antinuclear groups were not able to persuade Californians to pass a 1976 initiative to block further nuclear power development, they were able to scare the state legislature into passing draconian nuclear safety legislation that effectively accomplished the goal of the initiative—blocking any new nuclear power plants.[18] Now that nuclear power is less popular than it was during the energy-crisis years of the 1970s, potential investors in nuclear power doubtless recognize that states with the initiative process are especially poor candidates for new nuclear power plants.

To understand the role of public opinion, one must recognize that it is more than a matter of what the public as a whole thinks. Finding out what position a majority takes or which way opinion trends are moving is certainly useful, but it is only part of the story. When politicians and other decision makers take public opinion into account, they look beyond the bottom line of which side has majority support on an issue to examine subgroups, issue dynamics, and the details of how various people think about problems. They want to know details because some of those details have political consequences. Politicians, political activists, and decision makers want to know the answers to such questions as: Regardless of what the majority wants, what do likely voters want? Do people who live inland dislike offshore oil drilling as much as coastal residents? How much do people understand about energy technology, and what sorts of public information campaigns might change their minds? By exploring these and other questions ourselves, we will be able to gain an appreciation of the influence of public opinion on public policy.

SOME THEORETICAL QUESTIONS

Not all questions about public opinion are of interest to politicians or political activists. Some questions are of purely "academic" interest. In the following pages, I will pursue some of those questions in addition to those of more direct application in the world of politics and public policy.

A central academic question is, simply, what causes public opinion on environmental issues? Unlike most political issues of today, demographic variables such as education, income, age, and race only poorly explain what people think about environmental issues. To be sure, demographic variables do explain something. As we shall see, the young and the well educated tend to be more pro-environmental than the old and poorly educated. Still, age and education predict opinions on other social issues, such as abortion or civil rights, better than they predict environmental attitudes. Moreover, knowing a person's income gives an observer a far better guess at that person's opinion on tax issues than on any environmental question. In fact, as we shall see, income, race, and ethnicity—which divide American society on so many critical issues—are largely unrelated to opinions on energy and other environmental issues.

The prevailing theories about the causes of environmental opinions focus on deeply held values or cultural worldviews. Perhaps most prominently, anthropologist Mary Douglas and political scientist Aaron Wildavsky have developed "cultural theory" to explain opinions on environmental issues.[19] Their core argument is that deeply held cultural biases push "egalitarians" toward environmentalism because they can then use environmental issues to attack the existing capitalist system. In contrast, the cultural biases of "individualists" cause them to defend the system and take more pro-development stands. An alternative theory proposed by Ronald Inglehart maintains that recent generations who have grown up in relatively prosperous times have developed "postmaterialist" worldviews that cause them to value environmental purity more highly than do older generations who hold "materialist" worldviews.[20] My data on attitudes toward energy policy provide ample opportunity to test these theories and to improve on them.

THE DATA SOURCES FOR THIS STUDY

In this study, I use both national and California survey data to examine what people think about oil development and nuclear power. I rely more heavily on California data than national survey data because of the existence of a series of in-depth public opinion surveys of Californians conducted by the Field Institute,[21] and because Californians have been integrally involved in a series of major energy-policy struggles. The Santa Barbara oil spill helped trigger the modern environmental movement, and ever since California has a history of struggling with the most controversial oil development policies in the nation—those surrounding offshore oil drilling. California is also the site of several nuclear power plants, some of which have been targets of massive antinuclear protests. As a consequence, energy policy has gotten a great deal of public attention in the state.

California is not representative of the entire United States. Nevertheless, it is both large enough (12 percent of the U.S. population) and nearly enough representative so that we can reasonably generalize from California data to the entire United States.[22] For example, if the young tend to be far more pro-environmental than the elderly in California, it is a safe bet that the same pattern will also hold in the United States as a whole. As we shall see, where comparable national and California data exist, the patterns nicely match.

Finally, I should observe that California is important in its own right because of its size, because of its potential for further energy development, and because of its potential impact on the rest of the nation. Offshore oil development may not occur in the backyards of people living in Iowa, but decisions made by Californians affect the nation's petroleum supply and gasoline prices everywhere—even in Iowa City.

THE CONVENTIONAL ENERGY
FOCUS OF THIS STUDY

Environmentalists may be disappointed with this book because it focuses on public opinion toward conventional energy sources—oil development, nuclear power, and, to a lesser extent, coal. I largely ignore conservation and alternative energy sources. In doing so, I may seem to be encouraging development of conventional sources simply because I focus on them throughout the study. I urge readers not to jump to that conclusion.

Americans have faced a series of major policy decisions about oil development and nuclear power since the first energy shortages in the early 1970s. Although environmentalists have encouraged the nation to turn toward alternative, renewable energy sources, the public has not been faced with any major decisions on these matters. Almost all public policy decisions about alternative energy sources have been decisions about whether to pay for more research or whether to subsidize solar, wind, and other renewable energy sources. While these decisions are no doubt important, they have not dominated the nation's agenda, as have questions about oil and nuclear power. There have been many mass protests against offshore oil drilling and nuclear power, but how many protests have there been in favor of solar power or demanding tougher home-insulation standards? None.

Not only have oil and nuclear power questions dominated public discussion about energy policy, but they have also been the subjects of most public opinion surveys on energy issues as well. Pollsters generally ask questions only about issues that have been in the news because they have learned that people do not give useful answers to questions about obscure topics. When asked a survey question, a typical respondent will answer even if he or she knows little or nothing about the subject. Some researchers have explored this problem by asking survey questions about nonexistent laws and other fictional issues. Respondents have dutifully stated their opinions on such issues as the "Metallic Metals Act" and the "Public Affairs Act."[23] This is generally described as the "nonattitudes" problem.[24] Most people know too little about wind power, solar power, and other alternative energy sources to offer meaningful opinions on them. As a result, pollsters have asked few questions about them and I spend little time examining them. Again, I am not suggesting that alternative energy is unimportant, only that it has not captured the public's attention enough to warrant detailed investigation.

PLAN OF THE BOOK

I will begin my exploration of public opinion toward energy issues in chapter 2 with a brief history of America's energy problems. This will provide the historical context necessary to interpret the public opinion data later in the book. I go back many decades before the first public opinion polls on energy, however, because there are a few useful lessons about public opinion toward oil development that can be drawn from history, even without modern survey methods. In addition to offering a conventional, narrative, historical account, I also draw on data from the U.S. Energy Department's *Annual Energy Review* to offer a "history by the numbers" for the years since World War II. I believe this gives a far more detailed and accurate assessment of our energy situation than narrative accounts that one finds in other sources. Moreover, combining the historical account, the quantitative data, and findings in subsequent chapters provides a solid basis for the discussion of the future of public opinion and energy policy in chapter 6.

In chapter 3, I examine trends in public opinion on energy issues in both the United States and California, and I develop some statistical models to explain them. A number of previous studies have described public opinion trends, but there has been relatively little work modeling those trends. Moreover, most of the explanations have been no more than discussions of the data, pointing out historical events that might explain changes in the trends. I go beyond those sorts of explanations to develop and test formal models. I will also examine some popular hypotheses about potential causes of future opinion trends—for example, the claim that because the young are stronger supporters of environmentalism than the old, as the years go by the nation as a whole will become more pro-environment as older generations die off.

In chapter 4, I investigate a topic that has been left largely unexplored—what the public knows about environmental issues. Political leaders and activists commonly assume that the public is paying attention to their debates and understands the great struggles over environmental policy. Are they? In this chapter, I will show what the public knows and what it does not, and I will examine the extent to which they organize their opinions into recognizably consistent views on environmental issues.

In chapter 5, I examine support for oil and nuclear power development among various groups, and I seek to explain the causes of attitudes toward different energy sources by building and testing statistical models. In doing so, I investigate a range of causes, including demographics and measures of ideologies and values. Included in my investigation are tests of Mary Douglas and Aaron Wildavsky's "cultural theory" and of Ronald Inglehart's "postmaterialism theory." These two theories, which seek to explain environmental attitudes, have received a great deal of attention in the scholarly literature, yet they have not been subjected to an equal amount of careful empirical testing. I test these theories and assess the claims made by their proponents. Finally, I propose an expanded version of cultural theory, one that explains environmental attitudes far better than the original.

I conclude in chapter 6 with a discussion of some lessons we can draw about environmental policy and about public opinion in general. In that chapter, I address a question of central concern—Will public opinion be a force for or against constructive change in energy policy in the coming years?

NOTES

1. Patrick Lee, "Anger Flares as Oil Officials Defend Gasoline Price Hikes," *Los Angeles Times,* 26 April 1996; Patrick Lee, "Texaco Led Run-up in Southland Gas Prices," *Los Angeles Times,* 7 May 1996, A1.

2. "You Say Gas Prices Are Up? Let the Political Games Begin," *Los Angeles Times,* 1 May 1996, B8; Ronald Brownstein, "Democrats Reassert Role of Government in Marketplace," *Los Angeles Times,* 2 May 1996; Ralph Vartabedian, "Dipping into the Strategic Petroleum Reserve," *Los Angeles Times,* 9 May 1996, D1; Robert G. Beckel, "Presidential Politics Infuses Gas-Tax Debate," *Los Angeles Times,* 12 May 1996, M2; Michael Hirsh, "Getting All Pumped Up," *Newsweek,* 13 May 1996, 48; Janet Hook, "House Votes to Repeat 4.3-Cent Gas Tax Hike," *Los Angeles Times,* 22 May 1996, A1; Sam Fulwood III, "Senate Strikes Deal on Minimum Wage, Gas Tax," *Los Angeles Times,* 26 June 1996, A1.

3. Eugene A. Rosa and William R. Freudenburg, "Nuclear Power at the Crossroads," in *Public Reactions to Nuclear Power: Are There Critical Masses?* ed. William R. Freudenburg and Eugene A. Rosa (Boulder, Colo.: Westview, 1984), 18–19.

4. David E. Nye, *Consuming Power: A Social History of American Energies* (Cambridge, Mass.: MIT Press, 1998), chap. 7.

5. Because of the nature of the blowout and the state of technology in 1969, no one knows how much oil poured into the Santa Barbara Channel. Estimates range from 235,000 to 3.76 million gallons. For a variety of estimates, see David H. Davis, *Energy Politics,* 2d ed. (New York: St. Martin's, 1978), 75; and Robert Sollen, *An Ocean of Oil: A Century of Political Struggle over Petroleum off the California Coast* (Juneau, Ala.: Denali, 1998), 62.

6. Tom Wicker, *One of Us: Richard Nixon and the American Dream* (New York: Random House, 1991), 507–18.

7. Robert C. Paehlke, *Environmentalism and the Future of Progressive Politics* (New Haven, Conn.: Yale University Press, 1989), 76.

8. Lamont C. Hempel, *Environmental Governance: The Global Challenge* (Washington, D.C.: Island, 1996), chap. 2.

9. S. George Philander, *Is the Temperature Rising? The Uncertain Science of Global Warming* (Princeton, N.J.: Princeton University Press, 1998), chap. 13; John Houghton, *Global Warming: The Complete Briefing,* 2d ed. (Cambridge: Cambridge University Press, 1997).

10. See Amory B. Lovins, *World Energy Strategies* (Cambridge, Mass.: Ballinger, 1971); Amory B. Lovins, *Soft Energy Paths* (Cambridge, Mass.: Ballinger, 1977); Amory B. Lovins and L. Hunter Lovins, *Energy/War: Breaking the Nuclear Link* (New York: Harper and Row, 1980).

11. Walter A. Rosenbaum, *Environmental Politics and Policy,* 2d ed. (Washington, D.C.: Congressional Quarterly, 1991), 252.

12. Quoted in Michael Smith, "Advertising the Atom," in *Government and Environmental Politics: Essays on Historical Development since World War Two,* ed. Michael J. Lacey (Washington, D.C.: Wilson Center, 1989), 244.

13. See Benjamin Ginsburg and Alan Stone, eds, *Do Elections Matter?* 3d ed. (Armonk, N.Y.: Sharpe, 1996).

14. Joseph G. Morone and Edward J. Woodhouse, *The Demise of Nuclear Energy? Lessons for Democratic Control of Technology* (New Haven, Conn.: Yale University Press, 1989), 132–38.

15. Christopher H. Achen, "Measuring Representation," *American Journal of Political Science* 22 (1978): 475–510; R. Douglas Arnold, *The Logic of Congressional Action* (New Haven, Conn.: Yale University Press, 1990); John Kingdon, *Congressmen's Voting Decisions* (New York: Harper and Row, 1981); Warren E. Miller and Donald E. Stokes, "Constituency Influence in Congress," in *Elections and the Political Order,* ed. Angus Campbell, Philip E. Converse, Warren E. Miller, and Donald E. Stokes (New York: Wiley, 1966).

16. See Thomas Raymond Wellock, *Critical Masses: Opposition to Nuclear Power in California, 1958–1978* (Madison: University of Wisconsin Press, 1998); William R. Freudenburg and Eugene A. Rosa, eds., *Public Reactions to Nuclear Power: Are There Critical Masses?* (Boulder, Colo.: Westview, 1984); Morone and Woodhouse, *The Demise of Nuclear Energy?* chap. 5; William R. Freudenburg and Robert Gramling, *Oil in Troubled Waters: Perceptions, Politics, and the Battle over Offshore Oil Drilling* (Albany: State University of New York Press, 1994); Robert Gramling, *Oil on the Edge* (Albany: State University of New York Press, 1996); Robert Jay Wilder, *Listening to the Sea: The Politics of Improving Environmental Protection* (Pittsburgh, Pa.: University of Pittsburgh Press, 1998).

17. On current efforts to appease voters, see Carl Ingram and Nancy Vogel, "Legislature

Chapter 1

Oks San Diego Electric Relief Package," *Los Angeles Times*, 31 August 2000, A1; Nancy Vogel and Dan Morain, "Governor, Legislators Moving toward Bailout of Utilities," *Los Angeles Times,* 5 January 2001, A1; Jennifer Kerr, "Voters Would Block Power Rate Hike, Davis Warns Wall Street," *Santa Barbara News-Press*, 3 March 2001, A3.

18. James C. Williams, *Energy and the Making of Modern California* (Akron, Ohio: University of Akron Press, 1997), 307.

19. Mary Douglas and Aaron Wildavsky, *Risk and Culture* (Berkeley: University of California Press, 1982); Aaron Wildavsky, *The Rise of Radical Egalitarianism* (Washington, D.C.: American University Press, 1991).

20. Ronald Inglehart, *The Silent Revolution: Changing Values and Political Styles among Western Publics* (Princeton, N.J.: Princeton University Press, 1977); Ronald Inglehart, "Value Change in Industrial Societies," *American Political Science Review* 81 (December 1987): 1289–303; Ronald Inglehart, *Culture Shift in Advanced Industrial Society* (Princeton, N.J.: Princeton University Press, 1990).

21. The Field Institute is a nonpartisan, not-for-profit public opinion-research organization established by the Field Research Corporation (550 Kearny Street, Suite 900, San Francisco, California 94108). Data from the Field polls are archived at the University of California's UCDATA, located at the UC Berkeley campus. Neither of these organizations is responsible for the analysis or interpretation of the data appearing in this book.

22. U.S. Census Bureau, *Statistical Abstract of the United States, 1998*, 118th ed. (Washington, D.C.: Government Printing Office, 1998), 28.

23. Howard Schuman and Stanley Presser, *Questions and Answers in Attitude Surveys: Experiments on Question Form, Wording, and Context* (New York: Academic, 1981), chap. 4.

24. Philip E. Converse, "Attitudes and Nonattitudes: The Continuation of a Dialogue," in *The Quantitative Analysis of Social Problems,* ed. Edward Tufte (Reading, Mass.: Addison-Wesley, 1970).

2

A Brief History of America's Energy Problems

To understand public opinion about energy policy, one must know something about the history of America's changing energy situation and the government policies fashioned in response. Public opinion does not exist in a vacuum. The Middle Eastern wars, the energy shortages, the lines at gasoline stations, and the environmental disasters of the last four decades have all influenced what the public thinks about oil drilling, nuclear power, and other energy sources.

The history also sets the stage for a look into the future, which is the subject of the last chapter of the book. Knowing how the public responded to the last jump in gasoline prices or environmental disaster offers a good basis for predicting how it will respond the next time similar events happen. Looking at past trends in energy production and consumption also provides a basis for predicting likely patterns of energy production and consumption in years to come. Combining those two types of information, along with some of the findings from the next chapters, will allow me to offer some predictions about future public opinion on energy issues and about future energy policy.

In this chapter, I sketch out the history of America's energy situation. Because I use both U.S. and California public opinion survey data in later chapters, California receives more attention than it might in a history with a broadly national focus. The chapter begins with a few brief observations about the oil industry, and Californians' reaction to it, in the years before World War II. Although these years are not the focus of this study and there are no public opinion data on any energy issue from this period, a few events during those years can help us understand current public opinion. The next section offers a conventional history of America's energy situation since World War II. In the following section, I supplement the conventional history with discussion of a series of charts and graphs describing our energy consumption, production, prices, and other

13

data since 1949. In the final section, I discuss America's energy future. Together, these perspectives on history set the stage for the discussion of public opinion that comes in the following chapters.

CALIFORNIA AND THE EARLY
OIL INDUSTRY

The modern petroleum industry began with two events. In 1854, Abraham Gesner invented a process for distilling kerosene from petroleum, giving the world a liquid fuel that was far cheaper than whale oil. In 1859, Edwin Drake drilled the first successful oil well in Titus, Pennsylvania. After that, the race to find and develop oil fields was on.[1]

The Chumash Indians had discovered oil in and around what is now the Santa Barbara, California, area long before Europeans explorers arrived to document it. The Chumash had their own industry—building boats and making innumerable other useful or entertaining things with the easily found oil and asphalt. The invading Europeans destroyed the Chumash culture, but by the 1850s, the asphalt deposits west of Santa Barbara were being mined once again, this time for roofing and paving material—much of which was sold to San Francisco. The news of Edwin Drake's discovery spread rapidly, and by 1867 the first gusher had been drilled in Ojai, just east of Santa Barbara. The first profitable oil well in the state took a few more years; one was finally drilled in Los Angeles County in 1876.[2]

Although wildcatters searched for oil regularly in the Santa Barbara area, the industry did not take off until a large field was found in Summerland in 1890. Within months, the bluffs and beaches of Summerland and Carpinteria were covered with oil rigs. What had once been quiet coastal towns were now industrial eyesores to many. In 1896, as the available land was being used up, some drillers reasoned that if the oil was under the beach, it was probably under the ocean immediately offshore as well. That year the first offshore wells appeared, drilled from piers stretching out through the surf. By 1902, there were over four hundred offshore wells, some on piers extending over 1,200 feet out from shore.[3]

Assessing the public's reaction to the Santa Barbara oil boom is difficult, but there is clear evidence that at least some local residents actively worked against the spread of the oil industry from the very beginning. In fact, resistance to offshore oil drilling began the same year as the first offshore oil drilling operation in Summerland, 1896.[4]

An 1899 editorial in the *Santa Barbara Daily Press* called for resistance to a proposed oil exploration effort in the town of Santa Barbara itself. The editors wrote, "It would be an unfortunate disaster if the beach front near Santa Barbara's waterfront should be disfigured with the ugly derricks of oil wells. An attempt to force these unsightly creations upon the shore beyond Castle Point should be met by united resistance on the part of the people as a whole and the individual owners of adjoining property."[5]

Even the Chamber of Commerce joined the effort to keep drilling out of Santa Barbara.[6] The resistance worked. A mix of governmental and public pressure persuaded the drillers to stay out of Santa Barbara. Whether the activists were a vocal minority or representatives of the majority view is impossible to say, but they were certainly numerous enough or powerful enough to be successful.

Santa Barbarans were not the only people to denounce oil development along the Santa Barbara coast. In 1901, the editor of the *San Jose Mercury* wrote, "The whole face of the town is aslime with oil leakages. . . . If my impracticable spirit were to have its way, the whole beautiful region bordering the Santa Barbara Channel would be reserved as a universal playground, never to be marred by the progress of industrial development."[7]

Throughout the 1920s, there were several efforts to control offshore oil development. Money motivated most of the efforts. Both the state of California and local governments wanted tax revenue. In 1921 the state legislature passed the Mineral Leasing Act, which declared that the state owned the mineral rights in the coastal waters. The state sold leases and collected taxes on offshore oil for the next eight years. However, because of the extremely low royalty rate and public displeasure with the rapid expansion of offshore drilling, the state surveyor general began to deny all new offshore drilling permits in 1926. When the state supreme court overruled him in at the end of 1928 and ordered that permit sales be restarted, the state legislature stepped in and stopped everything with emergency legislation repealing the Mineral Leasing Act and calling a halt to further offshore leasing.[8] The legislature reversed itself two years later, but anti-oil forces managed to put the question to the California voters in the form of a referendum. In this first direct measure of public opinion on offshore oil development, the people voted no. Two later ballot initiatives that would have allowed new offshore drilling and slant-drilling into offshore oil fields were also defeated. The California public did not want any more drilling on their coast.[9]

A number of communities along the California coast also acted independently to limit oil development in their jurisdictions. In Santa Barbara, there were anti-oil protests in 1929 and a drive for a no-drilling sanctuary in the Santa Barbara Channel. At least one town, Redondo Beach, banned oil development by ballot initiative in 1935—another clear expression of public opinion.[10]

With the beginning of World War II, disputes over oil drilling vanished. The country needed oil to fight the war, and few people worried about the details of how it was obtained, even if it came from their backyards.

The events described here allow us to draw an important—if by now obvious—conclusion. Public sentiment against oil development did not begin with the environmental movement of the 1960s. People did not suddenly wake up then and discover the beauty of the California coast, nor did they begin to fear the risks of oil spills and pollution only with the massive 1969 oil spill in the Santa Barbara Channel. We cannot say what proportion of the public supported or opposed offshore oil development in the decades before, but we can say that there was significant opposition to oil development. There were protests. There were editorial statements against oil development by newspapers. There were actions by local government leaders to resist oil industry expansion. There were votes at both individual-city and state levels in which further oil development was blocked. Taken as a whole, the record shows a century-long history of substantial anti-oil sentiment in California.

This conclusion is certainly not original. Indeed, all the historians and researchers who wrote the books and articles I cite in this section were well aware of the anti-oil views of Californians. Yet when environmental writers discuss the modern environmental movement, they imply that public opinion changed and people became active on environmental issues at the same time. Many seem unaware of what occurred earlier in the century. The standard explanation is that Rachel Carson's *Silent Spring* and the 1969 Santa Barbara Channel oil spill sparked a new awareness. To some extent, they may have. Yet we should not exaggerate the influence of Carson's book or of the oil spill on public opinion.

What little public opinion data exist indicate that by the mid-1960s, the public overwhelmingly favored government action to clean up the air and water. Public opinion analyst William Mayer writes, "There is little doubt . . . that the *salience* of environmental issues increased enormously in the last half of the 1960s. But did the public's views about environmental *policies* undergo a similarly rapid transformation? The evidence on this point is not extensive, but the data that are available speak in the negative."[11]

In short, people were worried about environmental problems and favored government action to do something about them *before* the Santa Barbara oil spill. Indeed, the evidence suggests that at least on oil issues, some of the public held "modern" environmental views as far back as the late 1800s. The amount of attention environmental issues received and the government role in these issues may have changed enormously, but the dislike of oil development was clearly there for a large number of Californians. When the problems occurred in their own backyards or along their own coast—even when that coast was not very close to them—they often had what we would regard as modern, pro-environmental views.

FROM WORLD WAR II TO THE 1990s

The years following World War II were the golden age of energy for the United States. Energy was cheap, and inflation-adjusted prices steadily fell from the late 1940s through the 1960s. The cheap energy helped fuel America's postwar economic boom. The bright promise of nuclear energy made many believe that the supplies of energy would be limitless and that the cost would continue to fall, fueling the economy with ever-cheaper energy.

Yet even during this golden age, problems were developing. In the 1950s the United States began to import more energy than it exported. Most of this imported energy came in the form of oil from the Middle East. Indeed, the United States was not alone in looking to the Middle East for energy. The Marshall Plan to rebuild Europe after World War II pushed Western European nations to convert from coal to oil as their major source of energy. Yet by the mid-1950s Europe still relied predominantly on coal; oil accounted for only about 20 percent of its total energy consumption.[12] The United States imported only a small amount, and, more important, it could still increase its own production more than enough to do without imports entirely should the need arise. Our dependence on Arab oil still awaited us in the future. So it was that the first oil-related, Middle Eastern crisis had little impact on the American public.

In 1956, Egyptian dictator Gamal Abdel Nasser—an Arab nationalist who had risen to power with fiery denunciations of Western domination of Arab nations—declared that the Suez Canal belonged to Egypt and dispatched his troops to take over the largely undefended waterway. After a few months of negotiations, the British and French—who regarded the Suez Canal as theirs—together with the Israelis, who had heard more than enough of Nasser's talk of destroying Israel, launched an invasion to recapture the canal. Nasser could not prevent the combined might of Britain, France, and Israel from retaking the canal, but he could easily deny them the real prize—the shortcut to the Middle Eastern oil fields. Nasser ordered his soldiers to sink rock-filled ships in the canal, rendering it impassable and cutting off the oil that flowed through it. Egypt's allies compounded the damage; Syrian saboteurs damaged pumping stations on the pipeline from Iraq to the Mediterranean; other saboteurs shut down much of Kuwait's oil production. Saudi Arabia joined the effort, embargoing oil going to Britain and France. The oil flow from the Middle East to Europe, and especially to Britain and France, was sharply reduced.[13]

The oil shortages seriously threatened Europe, but not the United States. President Dwight Eisenhower, furious with the British and French for invading Egypt, refused them any assistance with their energy shortages until they gave up and pulled out of Egypt. Only then did American oil begin flowing across the Atlantic to help Europe through the winter of 1957.

Although the Suez Crisis damaged the economies of Europe and humiliated the British and the French, it had minimal impact on what the American public thought about oil or the energy policies of the United States. In 1956 there was plenty of excess oil on the American market, so oil prices only increased modestly—despite the "oil lift" to Europe. Indeed, in the aftermath of the crisis, the stepped-up oil production left the world with more available oil than before the crisis, causing prices to fall. Moreover, the oil industry responded to the crisis by beginning to develop larger and larger tankers—eventually "super tankers"— that could bypass the Suez Canal and take their oil around Africa to European and American markets. The political instability of the Middle East was recognized as a problem by policy makers, but it had little effect on American energy policy.

Following the Suez Crisis, world oil prices continued to fall. The declining prices caused American oil producers to call for quotas on imported oil to protect their profits. At first President Eisenhower urged voluntary quotas, preferring to let the market operate freely. But voluntary quotas failed when a recession struck in 1958. As a consequence, Eisenhower finally gave in and, acting under the provisions of the 1955 Trade Act, imposed import quotas on crude oil—oil imports could not exceed 12.5 percent of U.S. oil production.[14] The quotas created two markets—an American market, with artificially high prices, and the rest of the world, in which production was increasing faster than demand, resulting in falling prices. That system of quotas, with occasional adjustments, continued for fourteen years.[15]

The public response to the two-tiered pricing system illustrates an important lesson about public opinion on energy policy. People will get angry and demand that politicians solve their problems if something is being taken away from them, but their response is much more muted when something that they did not have before is *not* being given to them. As Douglas Arnold puts it, the responses to giving people benefits and taking benefits away from people are asymmetrical.[16] The two-tiered pricing system created by the oil import quotas kept the price of oil higher than it would have been without the quotas, but the price of gasoline did not go up. Because prices were stable, people did not feel that anything was being taken away from them. They were in fact being deprived of lower prices, but that did not draw much attention or complaint. Oil companies were profiting from the system, but few people seemed to consider that an important issue.

In 1960, responding to a unilateral cut in the price of oil imposed by Western oil companies, representatives of Iran, Iraq, Kuwait, Saudi Arabia, and Venezuela met in Baghdad to form the Organization of Petroleum Exporting Countries, OPEC. The goal of the organization was to reverse the decline in world oil prices by regulating production. Given that the founding members produced over 80 percent of the oil on the world market, their goal seemed realistic.[17] Other nations joined the new group, eventually resulting in a cartel of seven Arab and six non-Arab states that collectively produced almost all of the oil upon which the free world depended.

Despite OPEC's economic potential, however, the Western powers did not take it very seriously in the beginning. Although OPEC managed to achieve a few minor victories, internal conflicts weakened it. Political disputes among Arab nations coupled with the ever-present incentive for each nation to under-cut the prices of its rivals in order to increase market share undermined any demands OPEC might make. World oil prices continued to edge down.

In these quiet times, well before the first energy crisis, the commercial nuclear power industry began. Although nuclear power had its roots in the atomic bomb, the Atomic Energy Commission and government scientists had been pro-moting nuclear power as safe, clean, and efficient since the early 1950s. If the government's experimental nuclear reactors on military bases ever had any prob-lems, the public never heard about them. Environmental groups generally sup-ported nuclear power as a clean alternative to coal-fired power plants. Even David Brower, who would eventually leave the Sierra Club to form the antinu-clear "Friends of the Earth" in reaction to the Sierra Club's early pronuclear position, supported nuclear power in the 1950s.[18] So when California's Pacific, Gas, and Electric Company (PG&E) proposed the first commercial reactor at Bodega Bay, there was no outcry. The initial public hearings on the proposal drew few people, none of whom protested. The Sierra Club actually endorsed the proposal.[19]

Then the opposition began to form. At first, the opponents consisted of con-servationists and local people who did not want the beautiful bay just north of San Francisco to be ruined. Peace activists who feared that nuclear fuel could be converted into nuclear weapons and fall into the wrong hands soon joined them. The approval process dragged out, eventually coming to a halt when geologists discovered that the proposed site was only a thousand feet away from the San Andreas Fault.[20]

The failure of the first proposal for a commercial nuclear power plant was only a temporary hindrance to the industry. Other power plants were proposed and were eventually approved and constructed. In response to these proposals, however, antinuclear advocacy groups began to form. The Friends of the Earth was joined by the Union of Concerned Scientists in 1969 and by dozens of other groups, local and national, by the 1970s. As the number of nuclear power plants increased, so did the chorus of opposition. Eventually, in 1974, even the main-stream Sierra Club joined the opposition, formally condemning nuclear power.[21]

No measures of public opinion in these years are available, but as the 1960s progressed nuclear power, once an unquestioned product of brilliant scientists, became a controversial new energy source.

The next Middle Eastern oil crisis began like the first, with Gamal Abdel Nasser. In 1967, Nasser returned to his unfinished business of destroying Israel. He forced the United Nations observers out of the Suez Canal zone, initiated a naval blockade to stop shipping to and from Israel in the Gulf of Aqaba, and began sending troops and equipment to Jordan to prepare for an attack on Israel. Other Arab nations openly joined in the preparations.[22] In response, Israel launched a preemptive strike, beginning what became known as the Six-Day War. In a stunningly brief assault, the Israelis captured the Jordanian-controlled part of Jerusalem, the Sinai, the West Bank, and the Golan Heights.[23]

As the Arab armies collapsed before the Israeli onslaught, an oil embargo began, targeting the United States and Great Britain. Five nations—Algeria, Iraq, Kuwait, Libya, and Saudi Arabia—sharply curtailed their shipments to Israel's allies or ceased production entirely. The flow of Iranian oil also slowed, because Iraqi ship pilots refused to help guide Iranian oil tankers down the Shatt-al-Arab into the Persian Gulf. In a final blow to Europe, an unrelated Nigerian civil war caused the loss of another half-million barrels of oil a day from the world market.[24]

Although the situation may have seemed dire, it was not. When the embargo began, supplies on hand were large enough to make up the immediate shortfall. The United States, Venezuela, and other nations quickly increased their production, and within a month world leaders recognized that the embargo had failed. The embargo caused neither a sharp rise in oil prices nor any other significant impact on the American economy. The embargo officially ended three months after it started.[25]

For the American public, the Six-Day War was a gripping series of news stories from the Middle East, but the oil embargo's effect on the home front was barely noticeable. Although the price of oil on the spot market briefly rose, the price of a gallon of gasoline in 1967 increased only a fraction of a penny over the 1966 price. The war gave the public little reason to connect Middle Eastern conflicts with oil shortages or energy problems. If anything, the response to the embargo made policy makers even more confident of their ability to handle oil supply problems. The combination of available oil stocks, increased production, and more and larger tankers had overwhelmed the problem. The golden age of energy seemed to be getting better and better.

On the domestic front, however, unrecognized problems were developing. Policy makers understood the relatively simple matters of energy production and consumption. What they did not understand, and apparently did not anticipate at all, was the growing public demand for oil as the fuel of choice coupled with the rise of vehement political opposition to virtually every kind of energy production on American territory, including oil.

While few if any recognized it at the time, the beginning of the modern environmental movement helped set the stage for the energy crisis. As noted, many observers date the origins of the modern environmental movement to the 1962 publication of Rachel Carson's *Silent Spring*.[26] Although her book was not about pollution associated with the petroleum industry or other energy producers, it did affect them. *Silent Spring* crystallized the nation's desire for a cleaner, healthier environment and helped persuade people that the government ought to do something about it. People, as we have seen in California, had apparently been concerned with pollution for some time, but neither they nor their political representatives had seen the issue as a topic for government action. Carson helped change that.[27]

In the next few years, smog—previously of interest only in the Los Angeles area, where smog was notoriously bad—became the subject of media attention across the nation. Early air pollution legislation had left the matter as a state problem, but a smog crisis in New York City on Thanksgiving Day in 1966 and deteriorating air quality in other cities focused America's attention on dirty air and pushed the problem onto Congress's agenda.[28] Even before Congress acted, environmentalists were urging utility companies to convert from coal to cleaner oil or natural gas to generate electricity. The following year, Congress passed the Air Quality Act of 1967, which required states to establish air-quality and automobile-emissions standards and to enforce those standards.[29] Although the law was soon attacked as ineffective and far too soft on industry, it was Congress's first step toward replacing domestic coal with imported oil.[30] The goal of a healthier environment was helping to set the stage for the energy crisis.

Although we can look back with 20/20 hindsight and see that air pollution and the energy supply were linked, few people recognized it at the time. Opponents of clean-air regulation pointed out that the regulations would increase costs to various consumers and that those costs would be passed on to the public, but no one warned of an impending energy crisis. Even if someone had, it is extraordinarily unlikely that the public would have heard the warnings or paid any attention to them. As we will see in chapter 4, getting the public to learn even simple facts about the energy industry—such as the fact that the United States needs to import oil in order to meet its energy needs—is a difficult task. As a consequence, the public demanded cleaner air without recognizing how it would affect the energy supply. In this area, as in most, the public focused on one issue at a time.

At the same time that the demands for action on air pollution were building up, oil production—especially in offshore areas and in the newly discovered fields in Alaska—became politically controversial because of environmental risks. When the federal government decided to lease a tract for oil development off the California coast near Santa Barbara in 1966, and still more tracts in 1968, the decisions sparked large protests in the area, causing delays and eventually leading to smaller leases.[31] Notably, these protests took place *before* the Santa Barbara oil spill. Oil development in Alaska ran into similar obstacles. The decision to allow oil drilling at the huge new field discovered at Prudhoe Bay produced no opposition, but a proposed trans-Alaska pipeline to move the oil across Alaska to Prince William Sound drew resistance both from people who were worried about the risk of oil spills and from others who worried about the pipeline's effect on migration patterns of wildlife. Although there were few protests at the actual site of the proposed pipeline, there were many protests scattered across the lower forty-eight states and eventually legal actions seeking to block its construction.[32]

The 1969 blowout of the Union Oil Company's offshore oil drilling Platform A fundamentally changed the environmental movement. The blowout dumped between one and three million gallons of oil into the Santa Barbara Channel, creating an oil slick that covered 800 square miles of ocean and beach. Its effects spread far wider than the waters off the Santa Barbara coastline.[33] The resulting media coverage, including nightly television photos of thousands of dead fish and oil-coated birds, sparked a massive public reaction and forced President Richard Nixon and Congress into a series of actions that further increased U.S. dependence on imported oil.[34]

In terms of public opinion, what happened is not clear. We have no pre-1969 survey questions on opinions toward oil development, and few of any kind on environmental questions. Yet as observed in the previous section, less may have changed than is commonly believed. William Mayer's survey data on public opinion toward air and water pollution show that the public favored government action in the mid-1960s.[35] We also have the evidence from the years before World War II, discussed in the previous section—political activists had worked against oil development; protests had occurred; local ordinances had been passed to limit or block offshore oil development. A statewide election in 1931 had blocked further oil leasing. Was public opposition to oil development greater after the 1969 Santa Barbara Channel oil spill? Probably, but we cannot be sure, and if so we certainly do not know by how much.

We can, however, be sure about the behavior of politicians and political activists. Their behavior changed, and the changes were dramatic and important. Environmental issues were politicized. Interest groups formed or expanded, lobbying Congress for new laws. When pollsters investigated public opinion, they found support for these news laws—and Congress went along.

The first response from Congress was the passage of the National Environmental Policy Act of 1970 (NEPA), which established national environmental goals, created the Council on Environmental Quality within the Executive Office of the President, and—most important of all—established the "environmental impact statement" (EIS) process. The requirement that federal agencies must analyze the environmental impact of any significant act they take and guarantee that the actions will not damage the environment in any legally unacceptable way opened the doors for a host of lawsuits over environmental issues. Legal entanglements created by the EIS process slowed thousands of projects.[36] Among them was the trans-Alaska pipeline, which was critical for developing the newly discovered fields on Alaska's North Slope. Leasing of new tracts for oil exploration and development on the outer continental shelf was also slowed down, and a special five-year moratorium on further leasing in the Santa Barbara Channel was eventually needed to resolve some of those suits. All of this pushed America into a more precarious position.[37]

The second response was the passage of the Clean Air Act of 1970. Unlike the Air Quality Act, the Clean Air Act allowed no balancing of the risks of pollution and the costs to industry or other economic consequences. The law laid down hard deadlines and gave the federal government real enforcement powers. This time, many utilities were given no choice but to convert from coal to cleaner-burning oil and natural gas for generating electricity.[38]

Throughout the 1950s and 1960s, total U.S. energy consumption increased, and oil consumption led the rise. Total U.S. energy production rose as well, but not as fast as consumption. By the end of the 1960s, energy production finally began to level off, and domestic oil production actually began to fall. Rising imports filled the gap, so that by 1970 the United States imported about 30 percent of its oil. Although at first no one other than a few energy policy analysts noticed, in the 1960s America developed a serious energy dependence on foreign oil.

The problem of increasing foreign imports was not America's alone. The booming U.S. and world economies of the early 1970s used up most of the available oil production and tanker capacity. In earlier years there had been excess capacity, and that excess had provided an ample cushion against the petroleum boycotts of the 1956 Suez Crisis and the 1967 Six-Day War, but as the 1970s dawned that cushion no longer existed.[39]

The new decade brought more than a rapidly growing need for foreign oil; it also brought energy shortages. The United States simply could not produce or even import energy fast enough to meet the rapidly growing demand. As a result—and as an ominous warning of what was to come—brownouts hit the East Coast in the summer of 1970. Government officials asked the public to restrict its use of electricity during peak-use hours and openly spoke of rationing. The term "energy crisis" was heard for the first time.[40]

The energy situation worsened further because of a seemingly unrelated problem—inflation. President Nixon's response to rising inflation was to declare a wage and price freeze in 1971. But because the real price of energy was increasing on the world market, Nixon's solution had the effect of keeping energy prices artificially low and thus encouraging energy consumption while offering little incentive for conservation.[41] Insofar as the nation had an energy policy, it was making the situation worse.

Throughout the following months the energy situation continued to deteriorate. President Nixon loosened oil import quotas and finally abandoned them as the crisis mounted.[42] When a presidential plea to "exercise voluntary restraints" on price hikes replaced the price controls in January 1973, gasoline and other petroleum prices surged upward.[43] By summer, during the peak vacation-driving months, politicians and journalists were scrambling to explain the gasoline shortages.[44]

In 1973 once again, an Arab-Israeli war contributed to an oil crisis. By now, however, the world energy situation had changed enormously. This time, America was vulnerable. The golden age of energy finally ended.

On Yom Kippur, the most holy of all Jewish holidays, Egypt and Syria launched a surprise attack on Israel. At first the invasion seemed to be succeeding. Israel's armies were retreating, and, far worse, their military supplies were rapidly dwindling. The destruction of Israel suddenly seemed terrifyingly possible. Within a few days, Israel had to call on the United States for additional supplies and other aid. When the American cargo aircraft began arriving in broad daylight, openly signaling that the United States was siding with Israel, Arab leaders were furious.

They reacted first by instructing their oil ministers, who were meeting in Kuwait to negotiate new oil prices with oil company representatives, to increase the price of oil by 70 percent. Two days later, on October 17, the OPEC leaders increased the pressure by declaring a partial embargo. Not only would oil deliveries to the United States and other Israeli allies be cut back, but overall production would also be reduced—in a series of monthly 5 percent steps, so that nations could not move oil around to circumvent the embargo as they had done in 1956 and 1967. President Nixon, focusing on the war rather than on oil problems, now announced a major emergency military aid package for Israel. The Arab response to this new affront was a total embargo on oil shipments to the United States. America was cut off.[45]

The result of the embargo was a wave of price hikes and gasoline shortages across the United States. The price of oil shot up from three dollars a barrel to $11.65 in three months.[46] The price at the gas pump jumped as well, and worse, there was not enough gas to go around. For weeks the sight of cars lined up at gasoline stations was commonplace. Tempers flared, and in some places police had to maintain order at gas stations. Skyrocketing prices and lines at gas stations scared people and led to demands for action. Possibly for the first time, the American public was paying close attention to energy policy.

The call for action came from business leaders as well. The clean-air standards had caused them problems, and they wanted the standards rolled back. Utility companies wanted to use high-sulfur oil and coal to fire their generators. Automobile companies wanted to delay deadlines for emission standards. They all pointed to environmental regulations, claiming that they were too tough and contributed to America's energy woes.[47]

President Nixon and Congress responded with the National Energy Emergency Act to address the crisis, but they failed to reach a compromise over the basic elements of increasing production, weakening environmental standards, controlling prices, and imposing a windfall-profits tax on U.S. oil companies.[48] By January the embargo began to break up, however, as some Arab nations sent shiploads of oil to the West to profit from the high prices. Moreover, it became clear that the total embargo on the United States had never really worked; nations that were not boycotted could easily take deliveries of Arab oil shipments while they routed other oil to the United States. The expansion of the world's tanker fleet had created a worldwide oil market in which targeting specific nations for complete embargoes could not work.[49] On 18 March 1974, before any legislative action was taken, the embargo was finally lifted. The Arab nations began to pump oil again—but this time at far higher prices.

The 1973–1974 energy crisis shocked and confused America. Most Americans did not know that the United States had to import oil in order to meet its needs. In the wake of the embargo, most Americans were unclear about what had really happened and undecided whom to blame. In fact, most people were not even certain that an energy crisis existed.[50] Oil company profits had started rising in the early seventies as the energy market tightened. When the embargo began, profits shot up over 52 percent.[51] Members of Congress and other oil industry critics openly questioned whether the oil companies had manipulated the situation to their benefit. The Senate began hearings to investigate possible conspiracies.[52] The well-publicized controversies over what had happened, along with disbelief that America could actually run out of oil, led many people to think that the energy shortage was a fraud perpetrated by the oil companies to make more money.[53] A 1974 Roper poll asked, "Some people say there is a real shortage of gasoline and fuel oil because demand has outrun supply. Others say there really is not a shortage of gasoline and fuel oil and the big companies are holding it back for their own advantage. What do you think—that there is or is not a real shortage of gasoline and oil?" Seventy-three percent said there was no real shortage.[54] Another poll, conducted a year after the embargo had ended, found that only 54 percent of the public realized that the United States had to import oil to meet its energy needs.[55] Other polls agreed—the public did not understand what had happened. More important, no public consensus about what to do had emerged. No energy policy could command broad support.

This sequence of events, which I label the "energy crisis cycle," is important because essentially the same sequence was repeated in each later energy crisis.

- When the demand for energy rose, creating a tight supply, energy prices rose sharply—starting the energy crisis cycle.
- Along with increases in energy prices came large increases in the profits of energy producers.
- Politicians and interest group advocates criticized the energy industry for profiting on other people's misfortune, and charged them with manipulating prices to increase profits. Some critics even claimed that the energy industry had fabricated the energy crisis to increase profits.
- Most of the public believed the industry critics. They did not accept claims that the energy crisis was real, and so they felt justified in demanding that the government fix the problem without any cost to the public.
- In response to public demands, some politicians sought to protect the public from high prices with price controls or subsidies—steps that worsened the crisis, because they encouraged energy consumption in a time of shortages.
- Spokespeople for a wide range of business interests joined the debate with a chorus of requests to relax various environmental regulations in order to save energy. To them, the energy crisis was an opportunity to beat back environmental advances. Under those conditions, working out an effective set of energy policies became especially difficult.

In the aftermath of the energy crisis, much changed. The increase in the cost of imported oil resulted in a huge drain on the U.S. economy and in shifts of wealth among Americans as well. More important, it cut away at our sense of economic security and confidence.[56] We can now look back on 1973 as the year in which U.S. prosperity stopped its long post–World War II rise and leveled off. Median family income, adjusted for inflation, dipped, not to return to the 1973 level for another fourteen years.[57] The optimistic expectation that despite occasional bouts of recession and recovery Americans would continue to grow wealthier was not fulfilled. There were certainly recessions and recoveries in the next decade and a half, but for the next two decades, 1973 marked the high point of American prosperity. The OPEC oil boycott, of course, was not the only cause of economic stagnation, but it was a major contributing factor, and it changed the way Americans saw their country.[58]

In the years immediately following the 1973–1974 energy crisis, the nation's attention turned elsewhere. President Nixon spent the remainder of his term in office seeking to avoid impeachment and removal from office. High energy and gasoline prices continued, and energy problems continued to appear in the news, but little was accomplished. For all practical purposes, Watergate removed any chance of real progress in dealing with energy problems until Nixon was out of the way.

One of the more publicized energy controversies in the aftermath of the energy crisis was over whether to build special tanker ports so that oil companies could ship liquefied natural gas (LNG) across the Pacific from Indonesia and Alaska.[59] The question arose in 1973, when the Western LNG Terminal Company began planning for three LNG terminals near Point Conception at the western entrance to the Santa Barbara Channel, near Port Hueneme at the channel's eastern entrance, and in Los Angeles Harbor. At first the issue failed to gather much public interest, but when the Los Angeles Times reported that "LNG has a potential for danger that is almost unthinkable when spilled on water" and went on to discuss potential explosions and the possibility of lethal LNG leaks, environmentalists and journalists began focusing on the issue.[60]

The dispute continued for the next four years, waged in the legislative and bureaucratic hearing rooms in Washington, the state capitol in Sacramento, and cities near the proposed LNG terminal sites. Newspapers regularly covered the issue; two books were even written about it, Time Bomb and Frozen Fire.[61] According to the polls, the public supported building LNG terminals, but its support was somewhat anemic. A 1977 Field poll, for example, showed that Californians supported the idea of building more LNG terminals by a 47–33 percent plurality. But the same poll showed that other means of producing or importing more energy were even more popular. Proposals to drill more offshore oil wells, build more nuclear power plants, and build more facilities to import foreign oil all received greater support than building LNG terminals. Moreover, the public seemed a bit confused by this new idea of liquefied natural gas. Twenty percent of the respondents in the Field poll said they did not know enough to offer an opinion, an unusually high number.[62] (We will return to the question of how much the public understood about LNG and other energy issues in chapter 4.) At one point, it seemed that a deal made in the state legislature had guaranteed that Point Conception would be chosen, but falling natural gas prices and the long-running controversy took their toll. Eventually the deal was put on hold, and in 1986 the building permits were withdrawn. LNG terminals were no longer an issue in California.

President Gerald Ford attempted to make progress during his brief time in office, but his presidency was largely spent resisting the liberal-dominated Ninety-fourth Congress, elected in 1974. Ford sought to increase domestic production, largely by moving away from price controls; congressional Democrats, however, wanted to work toward energy conservation and to continue to protect

consumers by limiting prices and imposing windfall-profit taxes on oil companies. Eventually the two sides worked out a compromise, the 1975 Energy Policy
and Conservation Act, which gave each side part of what it wanted. The act
provided for gradual decontrol of oil prices over four years, energy efficiency
standards for electrical appliances, improved fuel efficiency standards for automobiles, and petroleum stockpiling for possible future emergencies (the Strategic Petroleum Reserve); it also gave the president the power to ration gasoline
in an emergency.[63] Although with the act America had begun to fashion an
energy policy, no one in the Congress or the administration regarded it as a
long-term solution to America's energy problems.

Jimmy Carter entered the White House with the goal of developing an effective national energy policy and setting America on its way to energy independence—the elusive goal of turning back the clock to the 1950s, when we had
minimal oil imports and could have done without any. Carter's first step was to
create the Department of Energy, bringing many of the scattered agencies with
control over various aspects of energy policy under one, centralized bureaucratic
roof. Although Congress quickly passed the bill creating the new cabinet department, President Carter's first major piece of legislation, it refused to give it full
powers or to create the "energy czar" for whom some had asked. The interior
secretary still held a substantial amount of control over energy production, and
the Environmental Protection Agency and the Nuclear Regulatory Agency still
remained independent. Moreover a new, quasi-independent agency, the Federal
Energy Regulatory Commission, was created within the Energy Department and
given partial control over oil and natural gas prices.

Carter's next step was to ask his new energy secretary, James Schlesinger, to
fashion a national energy policy. Carter announced the resulting proposals in a
nationally televised speech in April 1977, in which he declared that the energy
situation was "the moral equivalent of war."[64] The overall approach was that
the government would intervene in the economy to encourage conservation and
domestic energy production. Specifically, Carter proposed raising oil prices as
an incentive to conserve energy and to develop alternative forms of energy; he
proposed subsidizing research into new technology for both energy conservation
and alternative energy production; he proposed a variety of regulations encouraging or requiring energy efficiency; he proposed converting some electricity
generation from imported oil to domestic coal; and he offered a variety of mech-
anisms to ensure that the poor would not be unfairly burdened or harmed by his
energy program. All told, his national energy policy had 113 separate provisions,
affecting almost all Americans in one way or another.[65]

After eighteen months of struggle, a reduced, fragmented version of Carter's
energy program finally passed Congress and received a presidential signature.
Carter's proposals had run into trouble because of a variety of weaknesses. First

and most important, the president failed to persuade the public that the energy situation was, indeed, the moral equivalent of war. About half the public continued to doubt that the energy shortage was real; huge numbers still thought that the oil companies were conspiring to hold down supplies in order to drive up prices and profits.[66] Given the public's skepticism, it is small wonder that Carter never managed to persuade the public that the situation called for serious sacrifices. Carter's own label for the struggle was reduced to its acronym, MEOW, and made the butt of innumerable jokes. Second, Schlesinger's strategy of closeting himself with his energy experts and developing a perfect strategy for an ideal world ignored the task of building a political coalition—a basic necessity if the president was to get anything through Congress. Some of Schlesinger's proposals—most prominently, increasing the price of gasoline—simply had no support from either the public or the members of Congress who depended on it for reelection. The economists' argument that increasing the price of gasoline would help relieve the energy shortage by discouraging gasoline consumption may have been technically correct, but persuading the public to accept higher taxes is never easy. In this case, given that a large portion of the public believed that the energy crisis was fraudulent, persuading it to accept higher gas prices and other burdens was probably impossible. Third, many major organized interests would have suffered under Carter's plan. His proposals might have been good for the public in the long term, but it was not very nice to oil companies, automobile companies, utility companies, or a myriad of other powerful interests in the short term. The outcome was, in the words of one environmental advocate, the "functional equivalent of surrender in the moral equivalent of war."[67]

The second energy crisis began more quietly than the first, at least from the perspective of the American public. Throughout 1978, the Ayatollah Ruhollah Khomeini, a fundamentalist Shi'ite Islamic opponent of the shah of Iran, had been calling for increasingly violent demonstrations against the shah in an effort to topple him. In December, those demonstrations peaked in a month of violence that completely shut down Iranian oil exports. The following month, the shah abandoned his country to Khomeini and his followers.[68]

Most Americans did not foresee the consequences of the fall of the shah. Iran, after all, supplied only 5 percent of the world's oil, and Saudi Arabia announced that it would increase production to make up part of the shortfall. Yet the Iranian oil supplies contributed enough to the world supply to cause governments, oil companies, and consumers to begin worrying about a shortage and buying more oil to stock up. An oil panic set in. Between hoarding and speculative buying, prices climbed sharply throughout 1979 and into 1980.[69] In the fall of 1980, just as the world was finally beginning to adjust to the situation, Iraq added to the problem by invading Iran. The bloody Iraq-Iran War kept Iranian oil off the market and caused a 70 percent cutback in Iraqi oil exports as well.[70]

The results for America included higher prices, shortages, and the now familiar and infuriating lines at gasoline stations.[71] Prices eventually peaked in 1981 and then began sliding down, as demand fell during a worldwide recession.

As in the 1973–1974 oil crisis, oil company profits surged upward during the new crisis, and their critics cried foul.[72] The earlier debate about whom to blame for the shortages and high prices had never been resolved in the public's eyes. Following the 1973–1974 crisis, suspicions had eased, and the number believing that the crisis had been real grew somewhat, but the conspiratorial view of oil companies was still firmly established. Further, since the late 1960s Americans' confidence in government and business had been falling—and of all the industries and companies about which confidence-survey researchers asked, the oil industry and individual oil companies fared the worst after 1973.[73] Simply put, remarkably few people trusted them. When the second oil crisis struck, these issues were exacerbated. A huge percentage of the public still did not recognize the basic problem. A January 1979 Roper poll showed that 17 percent of the public admitted that they had no idea of how much of the oil used in the United States had to be imported, and another 41 percent substantially underestimated it. In March 1979, a CBS/*New York Times* poll asked, "Do you think the shortage of oil we hear about is real, or are we just being told there are shortages so oil companies can charge higher prices?" Sixty-nine percent responded that the public was just being told that shortages existed, and another 11 percent were not sure. Only 20 percent believed the shortages were real.[74] The public's overall distrust of oil companies and doubt that broad sacrifices were needed made dealing with the oil crisis especially difficult for politicians.

President Carter responded to the crisis with a new set of proposals, and this time he met with some success. He asked for subsidies for a synthetic fuels program and for a commitment from the Department of Defense to buy the new fuels, thereby creating an instant market for the new product. The goal was to produce substitutes for gasoline and natural gas from coal, biomass (such as cornhusks), oil shale, and other sources of domestic energy. Carter also declared that he would exercise his prerogative under existing price-control laws to decontrol the price of oil. Because this would result in a quick price increase and a surge of profits for oil companies, Carter asked for a windfall-profits tax. Both proposals won congressional approval. However, a third proposal, to create an Energy Mobilization Board with power to cut through environmental regulations to speed up energy production and conservation measures, died—the victim of environmentalists and consumer advocates who feared the disappearance of laws and regulations for which they had fought, and also of people who were worried about giving the White House such a huge grant of power. Despite this defeat, a set of policies designed to cope with the nation's energy problems was emerging.

The second energy crisis was not the only blow to America's energy situation in 1979. That was also the year in which nuclear power lost the last of its promise as a future source of cheap, limitless energy. Driven by environmental and safety concerns, public opinion had slowly been turning away from nuclear power since the late 1960s.[75] The pace of power plant construction was slowing, construction delays were lengthening, and some utilities were beginning to cancel their orders for new plants.[76] Yet in 1978 a majority of the public still favored the development of nuclear power. The Three Mile Island disaster changed all that.[77]

On 28 March, a combination of errors by designers, maintenance crews, and operators at the Three Mile Island nuclear power plant on the Susquehanna River outside Harrisburg, Pennsylvania, caused a potentially catastrophic partial meltdown of the reactor core.[78] The nation was transfixed as it watched the situation develop over the next ten days and wondered whether Pennsylvania was to be the site of a nuclear disaster.[79] The public's misunderstanding of what was happening almost certainly exaggerated fears. A national poll conducted shortly afterward asked, "From what you've heard or read, do you think a nuclear power plant accident could cause an atomic explosion with a mushroom-shaped cloud like the one at Hiroshima?" Only one-third of the public knew that the answer was no.[80] A chilling line in a just-released movie, *The China Syndrome*, summed up what many thought might be the potential outcome. Describing a fictional possible meltdown in California, actor Michael Douglas said that it could "render an area the size of Pennsylvania permanently uninhabitable."[81]

The nuclear reactor at Three Mile Island was stabilized, shut down, and eventually—years later—decontaminated. There was no explosion with a mushroom-shaped cloud. No one was injured or killed in the accident. Indeed, according to the presidential commission that investigated the accident, only a trivial amount of radioactivity leaked into the environment, and there were no adverse health effects.[82] Nevertheless, the accident spelled the end for growth in the nuclear power industry in the United States. In the immediate aftermath, public support for nuclear power fell sharply—from 50 percent in January to only 39 percent in April. It rose again the following year as energy prices increased, but as the analysis in chapter 3 shows, there was substantial, long-term damage to the image of nuclear power.[83] Vehement mass protests occurred at sites where new power plants were proposed or were under construction. Regulators imposed new, expensive conditions on builders of nuclear power plants. And no more nuclear power plants were ever ordered.[84]

The Three Mile Island accident had no immediate effect on the energy situation in 1979; the loss of the electricity produced by a single power plant had a negligible impact on the total U.S. energy supply. But the long-term impact was another blow to the goal of reducing oil imports. In 1979, nuclear power plants produced 3.5 percent of America's energy needs.[85] Because of power plants that were already nearing completion, that percentage increased in the following years. Yet because no new plants were being started, the contribution of nuclear energy to the total energy supply was less than it would have been. Eventually, as existing nuclear power plants age and must be closed, nuclear power will contribute less and less of the energy consumed in the United States. As a consequence, one part of Three Mile Island's legacy was an increased reliance on imported oil.

Ronald Reagan's election to the presidency in 1980 sharply changed the direction of U.S. energy policy. Rather than seeking ways for the government to intervene in the economy and manipulate it to attain the goal of energy independence, Reagan aggressively sought to reduce or eliminate government regulation of the energy market and to emphasize energy production over conservation and environmental protection. The only exception to Reagan's free-market approach was that he sought a hefty increase in the government's assistance to the nuclear power industry.

Reagan's first step as president was to end price controls on oil.[86] This step was largely symbolic because the price controls were due to expire before the end of the year, but it clearly conveyed the direction he intended to go. He followed this with an unsuccessful attempt to abolish the Department of Energy, and with successful efforts to cut funding for the government's synfuels program, to cancel various energy conservation programs, to decrease taxes on the oil industry, and to increase offshore oil development.[87] In fact, the Department of the Interior's original proposal called for leasing nearly one billion acres of offshore land, a staggering increase over the fifty-five million acres offered by President Carter's programs.[88] Although lawsuits and political controversy eventually limited Reagan's drive to increase offshore oil production as dramatically as he wanted, he nevertheless managed to spur domestic production.[89]

1981—the first year of Reagan's presidency—also marked the post–World War II peak in world oil prices. Driven by conservation efforts and declining demand because of recession, oil prices began a steady fall that lasted for five years. As a result, energy shortages moved off America's front pages during Reagan's presidency. There were no more energy crises, no more brownouts, and no more lines at gasoline stations. The news about energy was largely about environmental dangers or outright disasters.

Among the environmental dangers regularly appearing in the news were the flagrantly anti-environmental secretary of the interior, James Watt, and the Environmental Protection Agency administrator, Anne Gorsuch Burford. Watt's controversial efforts to open up Alaska and the outer continental shelf to more oil development received a good deal of attention in Reagan's first term. Watt's frequent attacks on environmentalists who opposed him and on the very idea of protecting the environment antagonized a good deal of the public, and according to some observers, may have contributed to the backlash against Reagan's policies and the rise of environmentalism during his tenure.[90]

The 1986 meltdown of one of the reactors at the Chernobyl nuclear power generator in the Soviet Union grabbed America's attention. Unlike the Three Mile Island accident, which released a minuscule fifteen curies of radioactivity into the atmosphere, the Chernobyl reactor exploded, releasing nearly 100 million curies and killing thousands of people. As events unfolded, experts declared that the designs of the Chernobyl reactor and American reactors were fundamentally different, that no such accident could ever happen with American reactors, and even that the Chernobyl disaster showed the wisdom of American nuclear engineers. But the nuclear industry critics saw the matter differently. Many of them became convinced that the experts had lied, that cataclysmic disasters were possible, and that Chernobyl proved it.[91] Whether Chernobyl had a major impact on public opinion was not clear to observers at the time (it is a question I take up in chapter 3), but that it reinvigorated antinuclear activists was quite clear. Faced with determined opposition, the nuclear industry remained moribund.

The other major energy news in 1986 was that Saudi Arabia decided to increase production, which resulted in an especially sharp drop in prices. In fact, some oil analysts refer to the 1986 price collapse as the "third oil shock"—the 1973–1974 and 1979–1980 crises being the first two.[92] Yet although the 1986 collapse may have had a substantial impact on the world economy, the American public barely seemed to notice it. To be sure, the collapse resulted in huge layoffs in the oil industry and badly damaged the economies of Texas and Louisiana, causing a good deal of anger and frustration in those states.[93] But as we have observed, there is a vast difference between how the public reacts when something is taken away and how it reacts when something is given to it. The lower energy prices were met with mild pleasure, not euphoria. The public seemed pleased, but largely indifferent.

The last energy-related problem that began to gain public attention during the Reagan years was global warming—the scientific theory that so-called greenhouse gases, such as carbon dioxide (CO_2), would raise the earth's temperature. This was not a new theory. The hypothesis that man-made carbon dioxide could warm up the planet, and the first estimates of the magnitude of the effect, had been published over a hundred years before.[94] Global warming had actually become an established topic of scientific inquiry by the 1960s. But it was not until the unusually hot summer of 1988, and the recognition that the 1980s had been the warmest decade on record, that global warming gained any serious journalistic or political interest.[95]

When journalists finally began reporting on global warming, public awareness grew. According to one poll, only 38 percent claimed that they had heard of global warming in 1981, a number that edged up to 41 percent by 1987. The surge of attention in the late 1980s, however, pushed the recognition level up to 86 percent by 1990.[96] But there is far more to learning about a problem like global warming than merely recognizing a name for it. Learning about any new, complicated set of scientific findings is difficult (as chapter 4 explains). The fact that scientific experts were disagreeing made learning especially difficult. Throughout the 1980s and 1990s, polls found that many people who claimed to know something about global warming confused it with pollution or ozone depletion.[97] As recently as 1997, a national poll found that more people believed that ozone depletion is a cause of global warming (it is not) than either driving automobiles or heating and cooling homes (they are).[98] Given the low level of the public's knowledge about global warming, it is not surprising that the public had yet to push for steps to mitigate the problem.

No sudden policy changes resulted from the new attention to global warming, yet bit by bit the idea of global warming crept into the public discourse over energy and environmental issues. In 1989, President George H. W. Bush proposed an international conference to consider global warming and related environmental issues. In 1990, a group of over seven hundred scientists received news coverage when they petitioned President Bush to take action to prevent global warming. Another, much larger group petitioned him again two years later.[99] In this fashion—through academic studies, international conferences, and the political activism of scientists and of environmentalists who shared their concerns—global warming became an established part of the discussion about energy policy.

The 1988 election of Bush to the White House and another Middle Eastern war changed the direction of America's energy policy once again. Bush, a man with a background in the Texas oil business, was much more open to government intervention in the energy market than President Reagan had been. Even before the first rumblings of the Persian Gulf War, Bush was laying plans for a new, comprehensive national energy strategy.[100] But before he could take action, America was hit by another oil-related disaster.

On March 24, 1989, the captain of the supertanker *Exxon Valdez* left port with a full cargo of oil and piloted his ship straight onto Bligh Reef in Port William Sound, Alaska. Eleven million gallons of crude oil poured out of the 1,400-foot-long ship. Despite the efforts of thousands of workers in cleanup crews, the oil spread 470 miles along the Alaskan coast in the next two months, fouling nearly every beach and inlet in its path. The oil not only destroyed the beauty of the coast but killed thousands of sea mammals and an estimated quarter of a million birds, and it devastated the economy and the way of life of the area's native Alaskans.[101]

Like the 1969 Santa Barbara oil spill, the *Exxon Valdez* disaster drew a huge amount of media attention. Television viewers once again saw oil-covered birds and dead fish washed up along the shore. The vivid images coming from Alaska introduced a new generation to the connection between the oil industry and environmental disaster. One national poll found that over three-quarters of the public knew that Exxon was the company responsible for the disaster. Another poll showed that over 70 percent of the public even knew roughly how big the spill was.[102] Given the generally low level of knowledge that people have about most political issues and events (see chapter 4), this is an astonishingly high level of awareness. Because of the lengthy cleanup efforts and the criminal and civil actions against Exxon, the disaster stayed in the news for years, reminding the public of the dangers of transporting oil by ship.

When Iraqi dictator Saddam Hussein ordered his army to invade Kuwait in August 1990, energy policy was back in the news and on Congress's agenda. The immediate effect of the invasion was a jump in prices. The invasion shut off the flow of Kuwait's oil to world markets. Moreover, the ideas that Iraq now controlled 20 percent of OPEC production and 20 percent of the world's known oil reserves, and that Saddam Hussein was accordingly a major influence on world oil markets worried western nations. But this oil crisis was unlike the earlier crises. The United States had pumped 600 million barrels of oil into its Strategic Petroleum Reserve, and most other major oil-consuming nations had built up their own emergency stockpiles. Moreover, Saudi Arabia, Venezuela, and other OPEC nations quickly increased their production, so that by December the oil shortages had disappeared. Prices began falling even before the United States launched its attack to liberate Kuwait. The overall result was a few months of tension and higher prices, but no brownouts or gasoline lines. In this respect, the crisis had a lot more in common with the Suez Crisis or the Six-Day War than with the oil crises of the 1970s.[103]

In January 1991 the United States and its allies launched their counterattack. The Persian Gulf War was short, brutal, and completely dominated by U.S. missiles and warplanes. Iraq's army, the third largest and most powerful in the world, collapsed before the combined might of the United States and its allies. But before they fled from Kuwait, Iraqi troops destroyed Kuwait's oil industry and set over 730 oil wells ablaze, leaving an environmental and economic disaster behind them. The fires would burn for over nine months before they were all finally put out.[104] In the conflict's final hours, President Bush decided not to pursue the war into Iraq. Fearing that too many U.S. casualties might result from a ground war in Iraq, Bush allowed Saddam Hussein to remain in control.

With the renewed recognition that U.S. oil supplies depended on the unstable Middle East, Congress set to work on a new energy package. The resulting legislation did not return to the heavy-handed government control of oil and natural gas prices of the past, but it embraced a range of government regulation and intervention in the economy, as well as a dose of deregulation. The simple free-market approach of President Reagan was abandoned.

The energy bill offered tax incentives and regulations to increase conservation and the use of renewable energy, and to spur the development of automobiles that used alternative fuels. In another boost for alternative fuels, the legislation required federal and state governments to buy cars that used the new fuels. The bill also included funding for research into ways both to reduce the consumption and to increase the production of energy. Perhaps most controversially, the bill eased the licensing process for nuclear power plants (although with no effect, because it resulted in no new orders for nuclear power plants). Finally, the legislation deregulated the wholesale electric power market, allowing producers to sell power to the highest bidder.[105] As with other decontrol efforts, the reasoning was that the increase in competition would increase efficiency and cut the price of electricity to consumers. What members of Congress did not realize was that this was the first step down the deregulation path that would help create an energy crisis a decade later.

Overall, this legislation had the most sweeping impact of any since the 1970s. Yet despite the progress, the bill left several topics alone—what to do about drilling for oil in the Arctic National Wildlife Refuge and along the coasts of the lower forty-eight states, whether to deregulate the retail electric power market, and whether to impose energy taxes to reduce consumption.

At the time, the public barely understood what was happening in Washington. Although the struggle over the legislation received a great deal of attention inside the Washington Beltway, it did not get much in the rest of the country. The bill was too technical, and from the public's point of view, there was no great issue involved. The bill did not offer the spectacle of an epic battle between environmentalists and big business; rather, it offered a series of technical fixes—hardly the sort of stuff that draws widespread interest.

Following the success of George Bush and the 102d Congress with energy policy, America turned its attention elsewhere. President Bill Clinton was more interested in health care reform in his first two years in office, and the new Republican majority had other interests when they took over Congress after the 1994 elections. Once again, in the absence of an energy crisis or a Middle Eastern war, questions about oil and energy policy left the front pages.

Only twice during the Clinton administration did energy problems receive much attention. In the spring of 1996, America's gasoline prices rose thirty cents per gallon in the space of four short months. Outraged drivers complained. The oil companies blamed the price hikes on the rising cost of imported oil, fires in ARCO and Shell refineries, and the extra costs of producing a new, cleaner-burning grade of gasoline that was mandated by environmental laws in many states. Critics of the oil companies declared that the rising oil prices and other explanations did not explain the entire price hike and that the oil companies were profiteering.[106] Because 1996 was a presidential election year, politicians of both parties leapt to the defense of the American driver, some calling for a cut in gasoline taxes and others calling on the president to release oil from the Strategic Petroleum Reserve.[107] In the end, both steps were taken. Prices stabilized and then fell. By the summer of 1996, inflation-adjusted gasoline prices had reached their lowest level since the government began collecting records shortly after World War II, and probably since the first gallon of gasoline was produced for the first car. Prices continued to move down throughout 1997, 1998, and into 1999. The energy crisis was in the distant past, and both the public and its elected leaders seemed completely unaware that there would be energy problems in their future.

Then in the early months of 2000 the energy-crisis cycle began again, this time because of an OPEC squeeze coupled with a booming U.S. economy. Hoping to drive prices up, OPEC cut production. At the same time, sales of sport utility vehicles (SUVs) skyrocketed, because of the nation's growing affluence. Nearly half of all new cars sold in the first nine months of 2000 were gas-guzzling SUVs, minivans, or pickup trucks.[108] The combination of a drop in OPEC production and the American public's move away from conservation was a costly one. From January to March, oil prices rose from twenty-four dollars a barrel to thirty-four. Prices fell briefly in April but began rising again in May, eventually topping thirty-six dollars a barrel by September, shortly before the presidential election.[109]

Along with soaring gas prices came greater profits for oil companies and charges that the industry was manipulating prices. Newspaper stories such as "Chevron Earnings Soar on Higher Oil, Gas Prices" and "U.S. Questions Refiners on Gas Prices" drove home the point to the public.[110] Because 2000 was a presidential election year, the candidates were inevitably drawn into the fray. Although Texas governor George W. Bush and Vice President Al Gore tried to focus their campaigns on education, abortion, health care, and other standard issues, the high price of gasoline drew more interest from voters than any other issue in 2000.[111] The candidates had to respond.

George Bush, a former oil man like his father, argued that the best way to lower prices was to produce more oil, and that the best way to do that would be to open Alaska's Arctic National Wildlife Refuge to oil drilling. He also joined congressional Republicans in a push to lift the 4.3-cent federal gas tax. Although cutting the gas tax was certainly popular with voters, it encouraged consumption, making the underlying supply problem even worse.[112]

Vice President Gore countered that the way to cut gasoline prices was to reduce energy use and to pressure major oil-exporting countries, such as Saudi

Arabia, to produce more oil for the world market. He insisted that relaxing environmental standards to make it easier to drill for oil in the United States would be a serious mistake, and he argued specifically that the Arctic National Wildlife Refuge should be kept closed to oil exploration. Gore also called on President Clinton to tap the nation's Strategic Petroleum Reserve to help push down prices. Following Gore's lead, Clinton ordered the release of 30 million barrels in September.[113] Like Governor Bush, Gore wanted to alleviate the problem as quickly as possible.

During the same months, business leaders and conservatives began complaining about overly restrictive environmental rules. Nuclear power advocates saw an opening and began arguing for new nuclear power plants, pointing out that they would produce no greenhouse gases and calling them therefore "clean" sources of energy. Other critics of environmental restraints called for opening up oil drilling both in Alaska and along the East and West Coasts. To them, the energy crisis was an opportunity to refight old battles and perhaps win them this time.[114]

As gasoline prices stabilized in the summer, a new energy issue hit the front pages—California was running out of electricity, and prices there were soaring. The crisis hit first in San Diego in August, then spread to the rest of California in the fall, and soon it began to push up energy prices across all the western states. The crisis had many causes. On the demand side, the soaring economy helped to cause shortages. High-tech firms and the Internet had led the charge. One study estimated that power demand in Silicon Valley had increased by 12 percent in the year leading up to the crisis. That may sound like a small problem, because Silicon Valley is only one, relatively small area on the southern end of San Francisco Bay. But a single, low-rise warehouse packed with specialized computers (known as "servers") to run the Internet—a "server farm"—can draw 150 to 200 megawatts, enough electricity to meet the needs of 150,000 to 200,000 homes.[115] From the power grid operators' point of view, building the Internet was just like building new cities with virtual people. The power drain was staggering.

Population growth also played a role. All of the western states grew in the 1990s, but only Montana had built enough power plants to keep pace with its growing population.[116] The combination of real population growth and high-tech demand quickly exhausted the excess capacity that states had possessed earlier, in the 1990s.

On the supply side, there were a host of problems. Few new power plants were being built in the West—because of local opposition wherever they were proposed, because of environmental group resistance, and because of lobbying by power companies to prevent rivals from building new plants. In fact, no new power plants had been built in California in the 1990s. In addition, because of environmental concerns, utilities were under pressure to convert from dirty coal to relatively clean natural gas, which increased the demand for natural gas and therefore its price as well.

The supply of natural gas became a problem because of its status as the cleanest of the fossil fuels. It produces the least pollution and the least greenhouse gas when burned. Throughout the 1990s, with the encouragement of environmental groups, power companies had converted more and more to natural gas. But the supply of gas—and just as important, the network of interstate pipelines to transport the gas from wellheads to consumers—had not kept pace with the demand. When a major pipeline exploded in New Mexico in August 2000, the supply dropped low enough to send prices surging upward. The high price of natural gas was a blow to consumers not just of natural gas but also of electricity, because of the heavy dependence of utilities on natural gas for power generators. When those prices were passed on to electricity consumers in San Diego, the energy crisis began.[117]

In California, Republican governor Pete Wilson had led a drive in 1996 to decontrol the retail power market so that individual consumers could choose companies from which they would buy their electricity. The decontrol legislation accomplished this by forcing the large, state-regulated monopolies that both produced power and sold it to customers to sell all their power-generating facilities.[118] The companies became retailers, purchasing electricity on the wholesale market and selling it to individual customers. The customers could buy from any retailer they liked. They could opt for the lowest rates or go with a company that offered somewhat more expensive "green" energy from renewable power sources. Supporters of decontrol expected that dozens of companies would rush into the state offering consumers cheaper energy in the new free market for electricity. It did not happen.

The state legislature had felt that the three big regulated firms that had supplied power to the state—Southern California Edison, Pacific Gas and Electric, and San Diego Gas and Electric—needed a temporary cushion to help them adapt to competition, so they gave the big three some advantages in the new market. If customers did not take the initiative to change power companies, they stayed with their original companies. In the confusing new market, few people decided to change. Another advantage given to the big three was that in order to change suppliers, moderate-sized and large customers had to buy expensive new meters. There were other financial advantages as well. To protect consumers during the period in which outside energy companies were supposed to be setting up shop in the state, the legislature capped retail prices for several years. The end result was that the "deregulated" system was rigged to make it difficult for outside retailers to enter. Few did. The outcome was very little competition and none of the anticipated benefits of a free market.[119]

The deregulation bill also left in place the system of regulations on high-power transmission lines. Because the regulations prevented transmission lines from making as much profit as other investments in the power industry, no new transmission lines were built. When the power crisis hit California in the summer and fall of 2000, the transmission lines became bottlenecks, making it impossible to get enough power to where it was needed.[120]

Another aspect of the deregulation system was that a decision by the California Public Utilities Commission forced energy retail companies to buy power on the spot market rather than enter into long-term energy contracts, as other states had done when deregulating their power markets. Because of the price caps set by the state legislature, when wholesale electricity prices skyrocketed, the retailers could not pass along the cost to their customers until the temporary rate caps expired. The crisis hit San Diego first because the law allowed San Diego Gas and Electric's cap to expire first. The company promptly passed along the cost of electricity to its customers, who started screaming at the state legislature. The other major retail companies were forced to sell their customers electricity at rates far below cost. Throughout the fall of 2000, their losses mounted, pushing them toward bankruptcy and eventually requiring a state bailout.[121]

Although the growing demand for power would eventually have overtaken supplies and triggered a crisis anyway, a combination of bad luck and accidents pushed California over the edge ahead of schedule and started the current energy crisis. The weather was unusually warm in California, which boosted the demand for power for air conditioners. The rainfall in the Pacific Northwest was below average. With less water flowing down rivers and through power dams, the Northwest had less power both for itself and to send to other states.[122] The New Mexico natural gas pipeline explosion reduced supplies and drove up the price of natural gas. Also, a fire in the control room of one of San Onofre's nuclear power reactors took 1,100 megawatts off line.[123] Collectively, these problems drove up electricity prices and eventually led to rolling blackouts throughout California. The energy crisis cycle had started again.

Once energy prices began to surge upward, so did the profits of the energy-producing companies. Major producers, such as Enron, Duke Energy, and Reliant Energy, saw their profits rise by anywhere from 30 to 130 percent and their stock prices shoot up as well.[124] Anger about high electricity prices coupled with evidence of windfall profits drew accusations of profiteering and price gouging from politicians and consumer advocates. Some even suggested that the energy crisis was a fraud, manufactured to make money.[125] As in previous energy crises, the public responded with suspicion and doubt. A *Los Angeles Times* poll of Californians conducted in January 2001 asked, "Do you personally believe there is an actual shortage of electricity in California, or not?" Fifty-four percent of the public said there was no shortage; only 36 percent believed it was real.[126]

Led by Governor Gray Davis, many politicians and activists insisted that electricity rates be capped and the public protected from higher prices. Shortly after the crisis began, the state legislature passed, and Governor Davis signed, a bill to help San Diego customers pay their electricity bills.[127] Consumer advocates demanded that rates be frozen and objected to any effort by the state to bail out the utilities.[128] Throughout the fall and into the winter of 2001, Governor Davis continued to insist that consumers had to be protected from rate hikes, despite rolling blackouts. At one point, he warned Wall Street not to push for rate increases, because the voters would block any such action.[129] For Davis and almost every other state legislator, there were two goals—solve the crisis and avoid taking the blame. Forcing voters to pay higher electricity bills was the last thing they wanted.[130]

At the same time, business and pro-development leaders used the crisis to call for rollbacks in environmental regulations and for expansion of oil drilling and nuclear power.[131] President Bush responded with a series of strong energy-development actions. He abandoned his campaign pledge to reduce carbon dioxide emissions from power plants. He rejected the Kyoto Protocol, negotiated by President Clinton to reduce greenhouse gases and fight global warming, but never ratified by the Senate. He called for the extraction of more coal to replace imported oil, and he continued his demand that the Arctic National Wildlife Refuge be opened to oil drilling. Other ideas, such as expanding nuclear power and easing smog rules, were also floated.[132]

Eventually, the state government did take steps to begin solving the energy crisis. The state Public Utilities Commission approved the largest rate hike in the state's history. Increasing the price of electricity reduced the financial pressure on the utilities and gave consumers a hefty incentive to conserve energy. The larger problem, however, remained—California and the West did not have enough power to prevent blackouts in the coming summer.[133]

AN ENERGY HISTORY BY THE NUMBERS

One way to sketch out the history of America's postwar energy situation is to describe the prominent events and government policies since World War II, as I did in the previous section. The weakness with that approach is that it skims over the numbers, offering only vague generalizations. An alternative approach is to look at America's postwar years using data on energy consumption, energy production, and other relevant trends. In order to fill in the picture above, I take that path now.

In addition to filling in the historical details, presenting these data prepares us for the discussion in chapter 6 of the future of public opinion and energy policy in the United States. The data on past energy consumption and production will give a sense of future consumption and production patterns. Moreover, the data offer indications of the size of the problems the nation will face and of possible solutions as well. For example, should Americans expect that developments in alternative energy sources will allow a smooth transition away from coal and oil? As we shall see, the data suggest that the answer is no.

Before turning to the data, a caveat is required. Like many sets of government data, these data lag somewhat behind events. The process of collecting and aggregating the data takes time; consequently, the figures do not include the recent months of the 2000–2001 energy crisis. Although the data miss our current situation as of this writing, they highlight a critical aspect of our current energy situation. In 1999, the latest for which we have statistics, U.S. energy prices were at an all-time low in real, inflation-adjusted dollars. Within two years, energy prices had shot up, and the United States was once again in the midst of an energy crisis. This instability resulted from our dependence on imported oil. So long as we must import oil to meet our energy needs, we are susceptible to sudden energy shocks.

Let me begin with an explanation of measures of energy. When making comparisons among different types of products, it helps to adopt a standard measure. When examining energy data, the standard measure is the British Thermal Unit, or Btu—the amount of energy required to warm one pound of water one degree Fahrenheit.[134] When looking at an entire nation's energy use, of course, the Btu is a ridiculously tiny amount. Consequently, most analysts speak of quadrillions (1×10^{15}) of Btu, or "quads" for short. A quad is a much more useful unit—roughly 293 billion kilowatt-hours.[135] Some of the data that follow are in quads.

We can divide energy sources into six basic categories: petroleum, natural gas, coal, nuclear electric power, hydroelectric power, and a residual category of other renewable energy sources. The first four of these are nonrenewable, in the sense that there is a finite amount of each fuel in the world. Once we use them up, they are gone (although at present rates, the world's known coal supply should last for 375 years).[136] The other types of energy are renewable, in that the world should continue to produce them as long as it exists. Hydroelectric power is the most widely used form of renewable power. Other forms of renewable energy include geothermal energy, wind energy, solar energy, biomass energy (converting organic material, such as wood or peat, into more easily used forms of energy, such as heat or electricity), and ocean energy (derived from the kinetic energy in tides or waves).[137]

Renewable energy sources are infinite, in the sense that the rivers, the wind, the sunshine, and the tides will last forever, so long as we are concerned. In a more practical sense, however, renewable energy sources are finite because there are only so many rivers that can be fitted with hydroelectric dams, only so many geothermal vents, and so forth. Moreover, with current technology, most types of renewable energy contribute only small amounts of the nation's total energy consumption. Renewable energy may save humanity in the long run, but for those of us now living, renewable energy will continue to be quite limited. Nonrenewable energy—including petroleum—will be essential for the rest of our lives.

Nonrenewable energy currently accounts for almost all of America's energy consumption. As figure 2.1 shows, in 1999 petroleum alone accounted for 37.7 quads, or 38.6 percent, of all energy the United States consumed. Natural gas, coal, and nuclear power accounted for another 53.9 percent. Renewable energy sources contributed only 7.5 percent of the total mix—with hydroelectric power providing the bulk of it. Geothermal, wind, and solar power combined contributed only 0.5 percent. Over the years, many people have hyped alternative energy, but the reality remains that alternative energy is still no more than the "half-percent solution." Aside from hydroelectric power, reliance on renewable energy lies far in the future.

Since 1949, the mix of energy sources that America uses has changed substantially. As figure 2.2 shows, throughout the 1950s and 1960s America consumed rapidly increasing amounts of oil and natural gas. In comparison, consumption of other sources of energy remained relatively flat. With the first energy crisis of 1973–1974, the pattern changed. Petroleum and natural gas consumption

Figure 2.1   Energy Consumption by Source, 1999

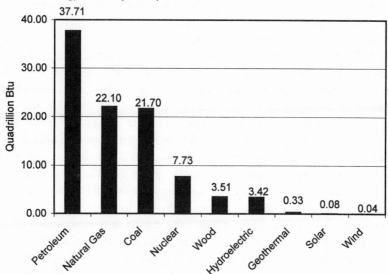

Figure 2.2   Energy Consumption by Source, 1949–1999

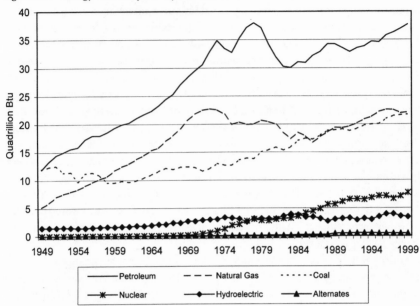

dropped, while coal and nuclear electric power contributed more to America's energy appetite. When the energy crisis ended, petroleum consumption quickly rebounded to a new high in 1978, but then fell again with the energy crisis of 1979–1980—a fall that continued during the recession that opened the 1980s.

Although the price hikes and recession accounted for most of the drop in oil use, the growing energy efficiency of the U.S. economy, prompted by the 1973–1974 crisis, also contributed to the decline. Most steps toward fuel efficiency—low-mileage cars and fuel-efficient generators, for instance—took several years to introduce to the market. So the second oil crisis hit just when America's first efforts toward energy efficiency were paying off.[138] Still, when the economy turned upward again, so did use of petroleum. The brief drop in consumption at decade's end, between 1989 and 1992, was also the result of the mild recession during the Bush presidency. By the mid-1990s, the United States was consuming almost as much petroleum as it ever had, and petroleum remained the dominant source of energy consumed by Americans.

One consequence of the first energy crisis can be seen in the surge of coal consumption. From 1949 to 1973, coal usage declined slightly and then edged back up. Although some observers liked to call America—particularly Montana and Wyoming—the Saudi Arabia of world coal, the coal resources of these states were largely untapped before 1973. Oil was the energy source of choice. OPEC changed that. Every president from Richard Nixon to Ronald Reagan sought to reduce American dependence on foreign oil by increasing use of domestic coal.[139] Their efforts—combined with the new price advantage of coal—worked. Many utility companies converted to coal-fired generators. America consumed so much more coal that in the mid-1980s, coal briefly became the second-largest source of the nation's energy.

The problem with coal is that it is a far dirtier form of energy than the oil or natural gas it replaces. Coal consumption has serious health and environmental costs. Some of the unfortunate side-effects of turning to coal, such as increased smog and other air and water pollutants, were well known long before the 1970s; others, such as acid rain and the possibility of a greenhouse effect, were only recognized later.[140] By the 1990s, many observers had come to think of coal as undesirable, but also unavoidable. In fact, the nuclear energy industry and its defenders began to argue that the possibility that America might have to use more and more coal in the coming years justified a reassessment of nuclear energy. Nuclear power might not be everyone's favorite option, they argued, but it was better than coal.[141]

Nuclear power, despite the growing public opposition since the 1960s and despite the lack of any new orders for nuclear power reactors after 1975, contributed a steadily increasing amount of energy from its earliest days until 1995, when production finally peaked. This growth resulted from a combination of nuclear power generators that were under construction finally coming on line, and of replacing the older equipment with newer, more efficient generators. Because no new nuclear power plants have been ordered for over twenty years, of course, nuclear reactors will contribute less and less toward the nation's energy needs as the older plants are decommissioned. In recent years, twenty-one plants have been decommissioned, representing a 17 percent reduction in nuclear electric power.[142] By 2015, about half of the existing nuclear power plants are scheduled to be shut down. Unless new nuclear generators replace them, the United States will not have any in operation by the year 2075.[143]

Finally, I should mention hydroelectric power. Although there has been some increase in the amount of hydroelectric power that Americans have used, it has not been very large. Unlike coal, there are no large, untapped reserves of waterpower. Utility companies had already built power dams in most of the practical places by 1973. Moreover, environmentalists were fiercely resisting efforts to build new dams by the time of the first energy crisis. Some new, small power dams have been built, and more efficient generators have replaced older, less efficient ones in existing power dams, but overall, the hydroelectric power industry is producing just about as much as can be expected from it. Indeed, many biologists and environmentalists are calling for removing some dams because of the ecological damage they do to rivers and the fish populations that depend on them. One hydroelectric power dam in Maine was removed in 1999 by federal order, and there have been discussions about dismantling more dams in order to restore the environment and rebuild salmon fisheries.[144]

Wind power, solar power, and other types of renewable energy also increased their contributions toward the nation's energy needs in the 1970s and afterward, but the amounts were relatively small. In 1977, when a Roper survey asked people whether they thought that solar, wind, and other power sources could realistically replace foreign oil within the next five years, 52 percent said they thought that solar power could do it, and 16 percent thought that wind power could.[145] These people had listened to too many exaggerated claims about the coming utopia; they were seriously mistaken.

The domestic energy production data presented in figure 2.3 begin to reveal the energy problem we face. The most obvious point to be gleaned from these data is the post-1970 decline in oil and natural gas production. The overall production of both fuels grew from 1949 until 1970 and then fell off—despite the energy crisis and the rising prices that oil and natural gas could command. The 1973 energy crisis brought brief, modest boosts in oil and gas production, but neither increase could be sustained. The 1986 collapse of world oil prices rendered some drilling ventures unprofitable and put domestic oil production on a

Figure 2.3   Energy Production by Source, 1949–1999

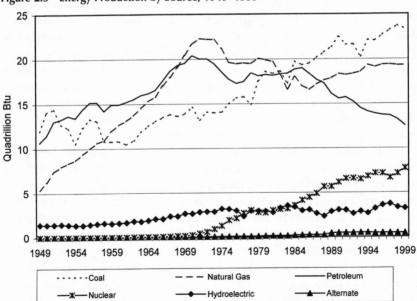

steady decline from which it has yet to recover. Indeed, since the mid-1980s oil production in the United States has dropped back to the level of the mid-1950s. Natural gas began rebounding in the mid-1980s, but is still well below its record production levels of the early 1970s. Simply put, America is running out of oil and gas. Production will never again return to those levels.

The growing disparity between consumption and production is shown in figure 2.4. Both consumption and production have been increasing over the last forty-five years, but consumption has been increasing more quickly. Imports fill the gap between the two. That gap is the heart of the energy problem. We consume more energy than we produce; consequently, we are at the mercy of the world's energy suppliers.

Figure 2.4 also shows why energy crises are not likely to just be events of the past. U.S. energy imports edged up throughout the 1960s and then began rapidly rising after 1970. The 1973–1974 OPEC oil embargo caused a brief decline in imports, but after 1975 energy consumption rose rapidly again, and the import gap widened. The 1979 energy crisis and the recession in Ronald Reagan's first years in office (the worst since the Great Depression of the 1930s) lowered imports back to the 1971 level, but when Saudi Arabia sharply increased production and the price of oil collapsed in 1986, imports began rising again. By 1997, America's thirst for foreign energy had surpassed its previous all-time

Figure 2.4 Total Energy Production and Consumption

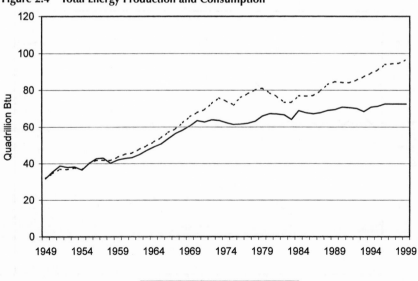

Production - - - - - Consumption

high in 1977. Crude oil imports reached a record high of 8.9 million barrels a day.[146] The only question was when OPEC would become sufficiently well organized to take advantage of the situation.

One might speculate that the rise in America's need for energy merely reflects the growing population. To a large extent, our growing population does drive our need for energy. But the data in figure 2.5 show that the amount of energy consumed by the average American has also increased enormously since the 1950s. Energy consumption per capita dropped after the two energy crises of the 1970s—driven down by higher energy prices, by efforts to reduce energy consumption, and by efforts to increase the efficiency with which we use energy. But the trend since 1982 has been upward, and Americans have nearly returned to the old heights of inefficiency at the start of the first energy crisis. The rising per capita use of energy results from two forces. On the one hand, efforts to conserve energy and use it more efficiently have had a huge impact. Steps ranging from lower-gas-mileage cars to better home insulation and more efficient electric motors have worked. Yet these steps toward fuel efficiency were largely driven by high fuel prices, which no longer exist, or government regulations, which were mostly passed in the 1970s and have already had their impact. The automobile industry, for example, has successfully lobbied against any further tightening of the CAFE fuel-efficiency standards. On the other hand, Ameri-

Figure 2.5 Energy Consumption per Capita

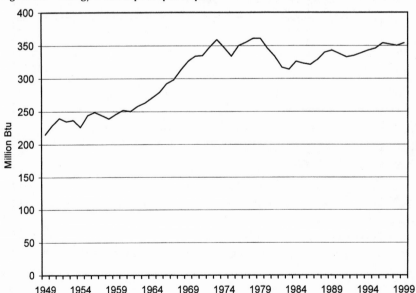

cans' standards have changed so that they now expect to have things they did not expect earlier—air conditioned cars and homes, for example. Moreover, we now use many energy-consuming products that were not widespread or did not even exist twenty years ago, from leaf blowers for the fall to snow blowers for the winter, to gasoline-powered lawn mowers for the summer (not to mention a bewildering array of other household and kitchen gadgets). Symbolizing this trend is the soaring popularity of the gas-guzzling SUV. Although it may not be as obvious as SUVs on the road, the Internet also sucks up an astonishing amount of power. The net effect is a return to our old energy-consuming ways of the past, but this time with fewer easy options for increasing energy efficiency when the need arises.

We can see the economic impact of our energy problems in several ways. We begin by looking at the prices of crude oil, gasoline, and heating oil in figure 2.6. These data show prices edging down very slowly from 1949 until 1973 and then suddenly jumping in response to the OPEC embargo. The next surge in prices started with the 1979 crisis and market panic, and continued in response to President Reagan's deregulation of oil prices. Eventually oil prices fell, driven down by the recession and Saudi Arabia's 1986 decision to increase production and create an oil glut (as noted, data for prices in the 2000–2001 energy crisis are not yet available).

Figure 2.6 Fossil-Fuel Production Prices in Real Dollars

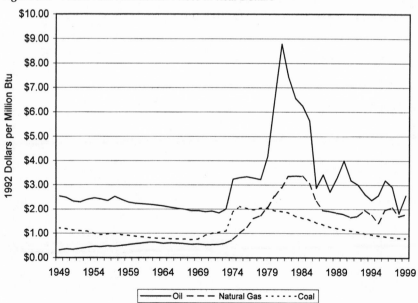

The low prices of the late 1980s contributed to the rising demand for energy. By 1992 the real, inflation-adjusted price of gasoline (including taxes) had actually fallen back to its pre-1973 level—despite the fact that gasoline taxes were higher and that clean-air regulations required most motorists to buy more expensive unleaded gasoline or—in some high-smog areas, such as Los Angeles, even pricier oxygenated gasoline. One of the best tools that we had to help solve our energy problems was the high price of gasoline, which served as an incentive for people to drive less and save energy. By 1992, that tool had been lost.

Although low petroleum prices had finally returned to America by the 1990s, the energy situation in the 1990s, as figure 2.7 shows, differed vastly from what it had been in the 1960s. What had once been a cheap, plentiful domestic product had become a cheap, plentiful imported product, and the cost of importing petroleum had become a substantial drain on the U.S. economy. Moreover, because of its dependence on imported oil, the United States had developed a national interest in maintaining stability in the Middle East. The Persian Gulf War stands as the most obvious evidence of this, but the war is neither the only evidence nor the only cost. One estimate of the annual cost of Gulf security in the mid-1980s was $40 billion. Once such costs are factored into the price per barrel, imported oil is a good deal more expensive than it initially appears.[147] On

Figure 2.7 Value of Fossil-Fuel Imports

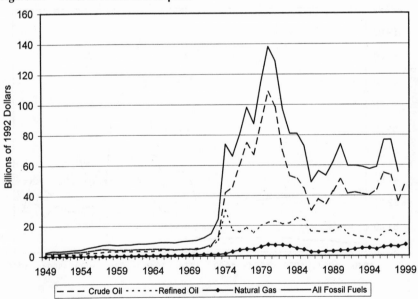

top of all that, because the United States imported so much oil, the situation was inherently unstable, as everyone learned in the new millennium.

OUR PRESENT SITUATION

In broad outline, we can divide the history of the U.S. energy situation since World War II into three periods, and we now seem to be entering into a fourth period. Following the war, the United States entered a golden age in which oil and energy issues were rarely mentioned, and then only as a series of positive messages—newly discovered Middle Eastern oil fields, continually declining prices, and the promise of nuclear power. In the late 1960s and early 1970s, the situation began deteriorating. The 1973–1974 energy crisis launched America into a second, more chaotic period in which prices shot up and America had to struggle with energy policy. From gasoline lines to laws mandating low-gas-mileage cars, Americans were given plenty of reasons to pay attention to energy policy. With the beginning of Ronald Reagan's first term as president in 1981, America entered a third period, in which energy prices fell and energy news mostly consisted of environmental threats and disasters. Although the energy policies of the 1970s left America with a host of little reminders in their day-to-day lives—gas-mileage stickers on new cars and energy-efficiency ratings for refrigerators, for example—energy issues were rarely debated in Washington. Questions about energy did not completely fall off the nation's agenda, but they certainly moved to the background.

Whether we are entering a fourth period of tighter energy supplies and higher prices is not yet clear. We are certainly in the midst of an energy crisis, but energy experts assure us that it will not last long. Supplies of natural gas and electricity should increase within a year or two, easing the power supply side of the problem. Nevertheless, we will remain at the mercy of the unstable regions that produce most of the world's oil. That instability may characterize the next period.

The third period—the 1980s and 1990s—was like the first in that energy prices were low (with the exception of the brief price spike of 1996) and in that our political leaders acted as if the long-term energy supply looked safe. Indeed, in 1999 energy prices had fallen well below the price levels of the golden age, which helps to explain why energy issues rarely made the news in those years. Americans had nothing about which to complain. But the situation fundamentally differed from the pre-1973 years in four important ways.

First, a large and growing percentage of America's energy needs were filled with imported oil. Unlike the 1950s, we no longer produce enough energy for our own needs. Given the declining productivity of our domestic oil industry and our growing energy consumption, it looks as if we never will again. At least as long as we depend on petroleum, energy independence is a thing of the past.

Second, a large and growing portion of the world's oil supplies came from one of the least stable areas in the world—the Middle East. Indeed, some petroleum experts believe that the Middle East may produce as much as 50 percent of the world's oil supplies by the year 2010.[148] With the likes of Saddam Hussein in power, that area is likely to remain unstable for the foreseeable future. In addition, the world's largest new oil fields are in another unstable area—the former Soviet states surrounding the Caspian Sea.[149] In short, we may not be having oil problems now, but smooth, uninterrupted supplies certainly cannot be guaranteed in the long term.

Third, scientists now understand far more about the environmental dangers of fossil fuels than they once did. In the 1950s, smog was regarded as an annoyance in a few cities, but not as a serious health threat. No one recognized any other major environmental heath threat. By the 1990s, scientists had learned about a wide range of risks that fossil fuels pose to both humans and their environment. We now know that toxins in the air can kill people, acid rain can kill forests, and a greenhouse effect may seriously damage the world economy and even—if the doomsayers are right—devastate the world. As a nation, we have taken steps to address many of these problems. We now have clean-air and clean-water laws, and we regulate a wide array of toxic substances, including many petroleum by-products. These regulations reflect the fact that the way we think about energy problems has changed because of our new understanding of the health and environmental costs of various policy choices since the 1950s. More generally, any steps we take to address our energy supply problems must now address the associated environmental problems as well.

Fourth, the American environmental movement has grown both in numbers and in political influence. What was once a relatively small group of people who focused mostly on preserving the wilderness and other conservation issues has now become a large number of environmental organizations—some with huge memberships—that focus on a wide range of issues running from protecting our air and water to preventing the construction of more nuclear power plants, to ending oil drilling off the California coast. Environmental organizations will be prominent players in the formation of any new energy policies in the coming years.

Taken together, these facts indicate that the 1990s were not another golden age of energy. They were not a repeat of the postwar years of plenty. If anything, they were no more than a temporary period of energy prosperity and stability—a temporary lull. We may once again return to a few years of low and seemingly stable energy prices. Yet given our dependence on foreign oil, we should recognize that such an idyll could end just as quickly as did the energy abundance of the late 1990s.

AMERICA'S ENERGY FUTURE

I now briefly look to the future and ask what it will bring. The central question here is, when will the world run out of petroleum? Or to put it less dramatically but more realistically, when will declining world oil production begin to affect the economy? This is an important question, because the answer will help us understand future public opinion and energy policy.

Asking about declining oil production is more useful than asking about the total world oil supply because there will be no single day when the world suddenly runs out of oil. Instead, there will be a gradual decline in oil production and a gradual recognition that the supply is dwindling and that we have to do something about it. To put it in more concrete terms, if the world has, say, a fifty-year supply of oil at the present rate of consumption, we may begin to feel the effects of oil running out in only another ten or twenty years.

The first consequence of diminishing oil production will be rising prices. Prices will rise because oil will become increasingly scarce, justifying higher prices. Prices will rise also because the cost of producing oil will increase. As the supply of easily produced oil runs out, oil companies will develop more expensive oil fields. In fact, this is already occurring. We no longer just produce oil by drilling in convenient places, such as Louisiana, Texas, and the Middle East. Oil companies now seek oil in remote places all around the world—from the frozen North Slope of Alaska to the ultra-deep-water wells in the Gulf of Mexico to the dangerous waters of the North Sea. The costs of producing oil in such places and of transporting it to market are high, but current oil prices justify those costs. As less and less oil remains, we will be producing more oil from high-cost wells, and the price of oil will rise accordingly.

As the price of oil rises, energy companies will move to bring other energy sources on line. These will surely include both conventional and alternative sources of energy. To some extent, the world can adjust to oil shortages by using more natural gas and coal, and by building more nuclear power plants—although these changes have significant environmental consequences. Wind, solar, and other renewable energy sources will also contribute more to the world's energy needs, both because technological advances should make them more efficient and because rising oil prices should make the higher prices of renewable energy more competitive. On top of that, as energy prices rise, people may consume less. One of the lessons of the energy crises of the 1970s, after all, is that demand for energy is elastic. As prices rise, consumption declines. At

some point, big cars may simply become too expensive, even if government regulations do not force downsizing.

The most feared potential consequence of declining oil supplies and rising
prices is a long-term economic downturn. It is easy casually to say that rising
energy prices will persuade energy companies to develop new technologies and
bring new sources of energy on line, but our technological prowess may not be
up to the challenge. Economists assume that supplies are determined by prices,
but what if our science and technology cannot fill the gap? The modern world,
after all, has never been without adequate petroleum supplies. We may not be
capable of making up the energy shortage at a price low enough to avoid long-
term damage to our economy. In addition to the possibility of technological
limitations, industry shortsightedness or unwise political decisions (urged by
poorly informed voters) may hinder progress toward filling the energy gap. In
short, the world does not work in the perfectly smooth fashion that some economic models like to assume. As our petroleum supplies dwindle, world economies may suffer.

Related to the problems of converting from oil to other sources of energy is
the problem of global warming. Although we do not yet fully understand global
warming, we know enough to recognize it as a serious potential threat. Many
observers argue that we know enough to follow the precautionary principle and
begin taking steps to ward off this threat. If global warming occurs to any substantial degree, it will seriously damage the economy.[150] In order to avoid it or
minimize it, however, we will have to reduce our use of fossil fuels and replace
them with energy sources that do not produce greenhouse gases—either nuclear
power or various renewable energy sources, such as wind and solar power. In
other words, if the global warming threat is real, we cannot turn to coal to
replace oil without suffering serious environmental and economic damage.

All of these issues turn on the question, when will declining world oil production begin to affect the economy? To state it differently, how long do we have
to prepare? Developing alternative sources of energy and conserving the energy
sources we have are not things that can be done overnight. The process will
take even longer if we need to shift away from fossil fuels to minimize global
warming.

HOW LONG WILL THE OIL LAST?

Although I leave questions about renewable energy and the speed of technological progress and improvements in energy efficiency to others, I will address the
central question of how long we have before energy supplies begin to dwindle
and the price of energy begins to rise. In broad outline, there are two approaches

to answering this question. First, one can count the amount of proven oil reserves, add a reasonable estimate of undiscovered oil, and compare that number to the rate at which the world consumes oil. Second, one can attempt to develop a model of the oil-discovery process and use it along with consumption data to predict the future.

Counting proven oil reserves and estimating the amount of likely future discoveries sounds like the easier and more accurate method; unfortunately, it suffers from the weakness that governments and oil companies manipulate, for political and business reasons, the numbers of "proven" oil reserves and estimates of undiscovered oil. There are only a few basic sources of information on world oil—*Oil and Gas Journal*, *BP Statistical Review of World Energy*, *World Oil*, the U.S. Geological Survey, and a private firm named Petroconsultants.[151] The data for the *Oil and Gas Journal* and the *BP* estimates come from an *Oil and Gas Journal* survey of governments and oil companies. These data are reported without any analytical efforts to correct errors (which will be discussed below). *World Oil* runs a similar survey, although the data differ from the *Oil and Gas Journal* data in some cases. Petroconsultants has developed a huge database of drilling, production, estimated reserves, and other useful information. Unlike the previously mentioned sources, rather than merely reporting data it has collected, Petroconsultants attempts to use the data to estimate reserves as accurately as possible. Even their data, however, may have errors, because governments and oil companies have reasons for misleading other governments and competitors. Finally, the USGS also assesses world oil reserves, largely depending on data from Petroconsultants.

For examples of likely inaccuracies with reported reserve estimates, consider the *Oil and Gas Journal* data discussed by oil geologist and consultant Collin Campbell, shown in figure 2.8. Campbell points out sudden jumps in reserve estimates and uses these examples to describe the games nations play with oil-reserve estimates:

> The first apparent anomaly was in Iraq in 1982, when an eleven billion barrel increase was announced, but in fact this was a delayed report of the discovery of the East Baghdad Field in 1979. The next anomaly was by Kuwait in 1984 when a 50 percent increase was announced without any discovery to justify it. Iraq accused Kuwait of exaggerating to secure a higher OPEC quota, which was partly based on *reserves*: apparently with good reason. Then in 1988, Venezuela decided to include about twenty billion barrels of heavy oil, which had been known for many years and was not in development. It had the not necessarily intended effect of increasing Venezuela's quota, and led Abu Dhabi, Dubai, Iran and Iraq to retaliate by announcing enormous unsubstantiated increases; followed two years later by Saudi Arabia.[152]

In other words, Campbell argues that petroleum reports are based both on the amount of oil believed to exist and on governments' economic and political

Figure 2.8   Unexplained Increases in Reported Oil Reserves

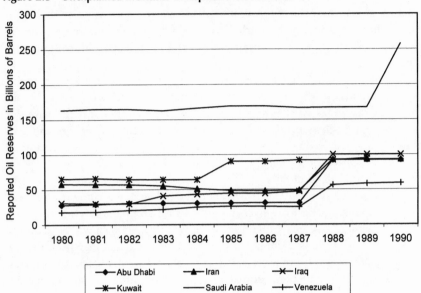

interests. Campbell also offers a number of other examples, such as Mexico's decision to increase its estimated reserves in order to demonstrate that it had the collateral to cover an international loan. Together, Campbell's examples and analysis make a persuasive case that the government and industry reports in the pages of *Oil and Gas Journal* exaggerate the amount of the world's oil.

If countries have incentives to exaggerate their reserves, firms have motives to misstate their oil holdings. Campbell observes that small oil companies with an eye on their stock prices tend to overestimate their reserves, while major firms may underestimate their holdings to avoid taxes and for other business-related reasons.[153] Taken together, these problems show that counting up proven oil reserves and estimating the amount of likely future discoveries is a crude method and one likely to overstate the world's oil supply.

The alternative to counting up oil reserves is to develop a model of the oil-discovery process and use it along with consumption data to predict the future. This is not an entirely different method, of course, because it relies on production data from the sources discussed above. However, production data are a good deal more reliable than estimates of reserves.

The basic model of oil production, first proposed by M. King Hubbert in 1956, assumes that both discovery and production of oil follow patterns that can be described by normal distributions over time, also called "bell-shaped curves" (see figure 2.9). In general, the model should apply to the production of any finite resource, such as oil, gold, silver, uranium, etc. The underlying insight is that once the resource is discovered, other producers join in the efforts to develop the resource, and the rates of discovery and production rapidly rise. Eventually, the rate of increase plateaus, when roughly half the existing resource has been discovered, and then falls as less and less of the resource remains. As the supply approaches zero, discovery and production efforts yield smaller and smaller amounts at greater cost.

When Hubbert initially published his model, he used U.S. production statistics to estimate the shape of the curve. On the basis of his results, he predicted that U.S. oil production would peak roughly in 1969 and then begin to decline. In the booming 1950s, few oil experts believed his predictions. Critics pointed out that Hubbert's model assumed that no new technology would alter the model's predictions. The normal-curve assumption implicitly presumed that new technology would increase oil production at a steady, predictable rate. By 1970, however, when U.S. oil production peaked on schedule (see figure 2.3), Hubbert's model and predictions were taken more seriously. Many oil analysts now use them in their own estimates of world oil supply and production.

These two approaches—counting up government and industry claims about their oil reserves, and relying on Hubbert's modeling methods—offer a range of predictions about how long the world's oil supply will last. We need not decide

Figure 2.9   The Rise and Fall of U.S. Oil Production

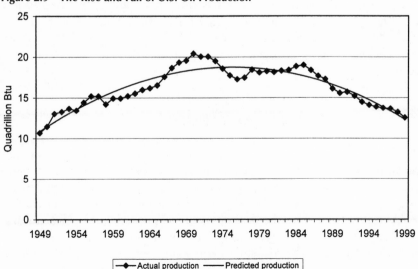

whose predictions are most accurate. For our purposes, merely summarizing the range of predictions is sufficient to make a general point about our future.

On the pessimistic side, some oil analysts see the peak of world oil production as almost upon us. Campbell predicts that oil production will hit its high point in 2001. This would mean that prices would start rising within perhaps another ten years and that the world would have enough oil to last another forty-five years at current rates of production and consumption. On the optimistic side, other analysts, relying on the *Oil and Gas Journal* figures, do not foresee world oil production peaking until perhaps as late as 2025, with enough oil to last perhaps another hundred years at current rates.[154] Other analysts offer estimates falling between those dates.[155]

These figures need to be qualified in two important respects. First, current production rates will certainly fall after the world passes its peak production, which will cause prices to rise. As noted above, the decline in production is virtually inevitable, so the idea of constant production for, say fifty or a hundred years, is just a way to think about the amount of oil remaining in the world—it is not a realistic prediction of future levels of production. Second, consumption rates will also change. In the short term, world petroleum consumption has been rising and will continue to rise because of the industrialization of China and third-world countries. In the long term, as oil becomes scarce and prices rise, consumption will fall. Again, describing the amount of oil remaining in the world in terms of current production and consumption rates is a good way to think about the amount of oil, but it is a nonsensical way to predict the future of the world's oil-dependent economies.

Given these estimates, we can say that rising prices and other consequences of declining oil supplies will begin to be felt some time in the next ten to forty years. To put it in more concrete terms, conversion from petroleum to other sources of energy is not a distant problem. It is a problem that we and our children will face.

CONCLUSION

As this brief history shows, many forces have influenced U.S. energy policy, and public opinion is one of them. We may not think of public opinion determining energy policy, but it has limited oil drilling and helped push the United States away from nuclear power. If energy were cheap and plentiful, those choices would seem obvious. Yet the days of cheap energy are limited. As petroleum production begins to diminish and prices begin to rise, the United States will once again have to make critical decisions about energy policy. When it does, public opinion will be a force to be reckoned with, for as we shall see, the public has definite preferences about how it wants its energy produced and what it wants to pay for it. The following chapters address those issues.

NOTES

1. Robert Gramling, *Oil on the Edge* (Albany: State University of New York Press), 13–14.

2. Robert Sollen, *An Ocean of Oil: A Century of Political Struggle over Petroleum off the California Coast* (Juneau, Ala.: Denali, 1998), 4–8.

3. Sollen, *An Ocean of Oil,* 9.

4. Paddock, Richard C. "Drilling Advance Rekindles Santa Barbara Oil Wars," *Los Angeles Times,* 5 December 1994.

5. Sollen, *An Ocean of Oil,* 11.

6. Harvey Molotch and William Freudenburg, eds., *Santa Barbara County: Two Paths,* Final Report to the Minerals Management Service, OCS Study MMS 96-0036 (Camarillo, Calif.: MMS, 1996); Robert Jay Wilder, *Listening to the Sea: The Politics of Improving Environmental Protection* (Pittsburgh, Pa.: University of Pittsburgh Press, 1998), 30–33.

7. Sollen, *An Ocean of Oil,* 11.

8. Wilder, *Listening to the Sea,* 34–37.

9. Ernest B. Bartley, *The Tidelands Oil Controversy: A Legal and Historical Analysis* (Austin: University of Texas Press, 1953), 71; Sollen, *An Ocean of Oil,* 12–17, 243.

10. Sollen, *An Ocean of Oil,* 17.

11. William G. Mayer, *The Changing American Mind: How and Why American Public Opinion Changed between 1960 and 1988* (Ann Arbor: University of Michigan Press, 1992), 102.

12. Daniel Yergin, *The Prize: The Epic Quest for Oil, Money & Power* (New York: Touchstone, 1991), 495.

13. For a fuller description of the Suez Crisis, see Yergin, *The Prize,* chap. 24, which is the basis for most of the description offered here.

14. David H. Davis, *Energy Politics,* 2d ed. (New York: St. Martin's, 1978), 80.

15. Yergin, *The Prize,* 535–40.

16. R. Douglas Arnold, *The Logic of Congressional Action* (New Haven, Conn.: Yale University Press, 1990), 32.

17. Yergin, *The Prize,* 523.

18. Thomas Raymond Wellock, *Critical Masses: Opposition to Nuclear Power in California, 1958–1978* (Madison: University of Wisconsin Press, 1998), 74.

19. Wellock, *Critical Masses,* 29–30, 68–72.

20. Wellock, *Critical Masses,* chap. 1.

21. Wellock, *Critical Masses,* 71.

22. "Middle East: The Scent of War," *Newsweek,* 5 June 1967, 40–48; "Intermission: 'Too Late and Too Early,'" *Newsweek,* 12 June 1967, 38–40; Yergin, *The Prize,* 554.

23. "Terrible Swift Sword," *Newsweek,* 19 June 1967, 24–34.

24. Yergin, *The Prize,* 555–56.

25. "Shadow War on the Economic Front," *Newsweek,* 19 June 1967, 68–70; "Oil: When Is a Ban Not a Ban?" *Newsweek,* 26 June 1967, 57, 59; Yergin, *The Prize,* 557.

26. Rachel Carson, *Silent Spring* (Boston: Houghton Mifflin, 1962).

27. Michael P. Cohen, *The History of the Sierra Club, 1892–1970* (San Francisco: Sierra Club, 1988), 285–91, 334–35; Charles O. Jones, *Clean Air: The Policies and Politics of Pollution Control* (Pittsburgh, Pa.: University of Pittsburgh Press, 1975), chap. 5; Mayer, *The Changing American Mind*, 102–106.

28. Jones, *Clean Air*, chap. 3; Harold W. Kennedy and Martin E. Weekes, "Control of Automobile Emissions—California Experience and the Federal Legislation," in *Air Pollution Control*, ed. Clark C. Havighurst (Dobbs Ferry, N.Y.: Oceana, 1969), 101–18; "Pollution Alert," *Newsweek*, 5 December 1966, 66–67.

29. Jones, *Clean Air*.

30. John C. Esposito et al., *Vanishing Air* (New York: Pantheon, 1970), 270–71; John E. O'Fallon, "Deficiencies in the Air Quality Act of 1967," in *Air Pollution Control*, ed. Havighurst, 79–100.

31. Gramling, *Oil on the Edge*, 89; Carol E. Steinhart and John S. Steinhart, *Blowout: A Case Study of the Santa Barbara Oil Spill* (Belmont, Calif.: Duxbury, 1972).

32. Mary Clay Berry, *The Alaska Pipeline: The Politics of Oil and Native Land Claims* (Bloomington: Indiana University Press, 1975); Charles J. Cicchetti, *Alaskan Oil: Alternative Routes and Markets* (Washington, D.C.: Resources for the Future, 1972).

33. As noted in chapter 1, estimates range from 235,000 to 3.76 million gallons. See Davis, 75; and Sollen, *An Ocean of Oil*, 62.

34. Harvey Molotch, "Oil in Santa Barbara and Power in America," *Sociological Inquiry*, 40 (Winter 1970): 131–44; Tom Wicker, *One of Us: Richard Nixon and the American Dream* (New York: Random House, 1991), 507–18.

35. Mayer, *The Changing American Mind*, 102–103.

36. Walter A. Rosenbaum, *Environmental Politics and Policy*, 2d ed. (Washington, D.C.: Congressional Quarterly, 1991), 92–96.

37. James Everett Katz, *Congress and National Energy Policy* (New Brunswick, N.J.: Transaction, 1984), 14.

38. Marc K. Landy, Marc J. Roberts, and Stephen R. Thomas, *The Environmental Protection Agency*, expanded ed. (New York: Oxford University Press, 1994), 28–30.

39. Katz, *Congress and National Energy Policy*, 13.

40. "Warning: Low Voltage," *Newsweek*, 18 May 1970, 123; "There Goes the Power," *Newsweek*, 10 August 1970, 65; "Power: Scrounging for Fuel," *Newsweek*, 14 September 1970, 89.

41. For an excellent short description of the politics of energy during the Nixon administration, see Katz, *Congress and National Energy Policy*, chaps. 3–4.

42. Yergin, *The Prize*, 567.

43. Helmuut A. Merkleim and W. Carey Hardy, *Energy Economics* (Houston: Gulf, 1977), 182.

44. Robert A. Rosenblatt, "Energy Crisis: Oil Firms, U.S. Caused Shortage," *Los Angeles Times*, 17 July 1973.

45. Yergin, *The Prize*, 599–609.

46. Congressional Quarterly, *Energy & Environment: The Unfinished Business* (Washington, D.C.: Congressional Quarterly, 1985), 19.

47. "Facing Up to Cold Reality," *Newsweek*, 19 November 1973, 109–10; "Next, the Oil Recession?" *Newsweek*, 3 December 1973, 87–90; "Up the Wall," *Newsweek*, 18 March 1974, 99–100.

48. For a description of the politics of this and later congressional efforts to deal with energy problems, see Katz, *Congress and National Energy Policy.*

49. Thomas H. Lee, Ben C. Ball, Jr., and Richard D. Tabors, *Energy Aftermath* (Boston: Harvard Business School Press, 1990), 30–37.

50. Roobina Ohanian Tashchian and Mark E. Slama, "Survey Data on Attitudes and Behaviors Relevant to Energy: Implications for Policy," in *Families and the Energy Transition,* eds. John Byrne, David A. Schulz, and Marvin B. Sussman (New York: Haworth, 1985), 29–51; Al Richman, "The Polls: Public Attitudes towards the Energy Crisis," *Public Opinion Quarterly* 43 (1979): 576–85; P. T. Thompson and J. MacTavish, *Energy Problems: Public Beliefs, Attitudes and Behaviors* (Allendale, Mich.: Grand Valley State College, Urban and Environmental Studies Institute, 1976).

51. "Top Oil Firms' 1973 Profits Jump 52.7%," *Oil and Gas Journal* 72 (18 February 1974): 32–34; "Oil's High Profits Abroad Bring Trouble in the U.S.," *Oil and Gas Journal* 72 (6 May 1974): 93–96.

52. Yergin, *The Prize,* 656–59.

53. W. H. Cunningham and B. Joseph, "Energy Conservation, Price Increases, and Payback Periods," in *Advances in Consumer Research,* vol. 5, ed. H. H. Keith (Ann Arbor, Mich.: Association for Consumer Research, 1978), 201–205; Louis Harris, "Energy Shortages Regarded as Serious by Most Americans," *The Harris Survey* (New York: Chicago Tribune, 26 July 1973); Louis Harris, "Energy Crisis Has Many Causes," *The Harris Survey* (New York: Chicago Tribune, 11 August 1977); "Energy: No Shortage of Suspicions," *Newsweek,* 14 January 1974, 63–64.

54. Richman, "The Polls," 577.

55. Michael X. Delli Carpini and Scott Keeter, *What Americans Know about Politics and Why It Matters* (New Haven, Conn.: Yale University Press, 1996), 320.

56. Lester Thurow, *The Zero-Sum Economy: Distribution and the Possibilities for Economic Change* (New York: Penguin, 1980), 26–31.

57. Kevin Phillips, *Boiling Point: Republicans, Democrats, and the Decline of Middle-Class Prosperity* (New York: Random House, 1993), 19–21; Harold W. Stanley and Richard G. Niemi, *Vital Statistics for American Politics,* 3d ed. (Washington, D.C.: Congressional Quarterly, 1992), table 12-2.

58. Although official government estimates of personal income and gross domestic product per capita show that the economy peaked in 1973, those estimates depend on the consumer price index, which many economists now believed to be flawed. If one takes the arguably exaggerated CPI into account, the economy and median family income did, in fact, grow after 1973. Yet even with these new estimates, the growth rate is estimated to be well below the growth rate from the end of World War II to 1973. See John M. Berry and Eric Pianin, "Panel Finds Price Index as Excessive," *Washington Post,* 5 December 1996.

59. William R. Ahern, "California Meets the LNG Terminal," *Coastal Zone Management Journal* 7 (1980): 185–221; Sollen, *An Ocean of Oil,* chap. 7.

60. Ed Meagher, "Liquefied Natural Gas—Risk of a Disaster Feared," *Los Angeles Times,* 12 January 1975.

61. Peter van der Linde with Naomi A. Hintze, *Time Bomb: The Truth about Our Newest and Most Dangerous Energy Source* (Garden City, N.Y.: Doubleday, 1978); Lee N. Davis, *Frozen Fire* (San Francisco: Friends of the Earth, 1979).

62. Data from Field Poll 7703. See the data appendix for details.

63. Katz, *Congress and National Energy Policy,* chap 4.

64. Steve Isser, *The Economics and Politics of the United States Oil Industry, 1920–1990* (New York: Garland, 1996), 146.

65. Katz, *Congress and National Energy Policy,* 99.

66. Richman, "The Polls," 576–85.

67. *New York Times,* 10 November 1978, D14.

68. Yergin, *The Prize,* 674–83.

69. George Philip, *The Political Economy of International Oil* (Edinburgh: Edinburgh, 1994), 162–63.

70. Edward T. Dowling and Francis Gittilton, "Oil in the 1980s: An OECD Perspective," in *The Oil Market in the 1980s: A Decade of Decline,* eds. Siamack Shojai and Bernard S. Katz (New York: Praeger, 1992), chap. 6; Yergin, *The Prize,* 706–14.

71. Katz, *Congress and National Energy Policy,* chap 8.

72. "Oil's High Profits Abroad Bring Trouble in the U.S.," 93–96.

73. Seymour Martin Lipset and William Schneider, *The Confidence Gap: Business, Labor, and Government in the Public Mind* (Baltimore: Johns Hopkins University Press, 1987), chap. 1.

74. Richman, "The Polls," 577–79.

75. "Power: Sluggish Atom," *Newsweek,* 23 June 1969, 81–82.

76. Joseph G. Morone and Edward J. Woodhouse, *The Demise of Nuclear Energy? Lessons for Democratic Control of Technology* (New Haven, Conn.: Yale University Press, 1989), chap. 5.

77. Robert C. Mitchell, "Public Responses to a Major Failure of a Controversial Technology," in *Accident at Three Mile Island: The Human Dimensions,* ed. David L. Sills et al. (Boulder, Colo.: Westview, 1982), 21–38; William R. Freudenburg and Eugene A. Rosa, eds., *Public Reactions to Nuclear Power: Are There Critical Masses?* (Boulder, Colo.: Westview, 1984).

78. For a fuller description, see Joseph G. Morone and Edward J. Woodhouse, *Averting Catastrophe: Strategies for Regulating Risky Technologies* (Berkeley: University of California Press, 1986), 48–55.

79. Peter S. Houts, Paul D. Cleary, and Teh-Wei Hu, *The Three Mile Island Crisis: Psychological, Social, and Economic Impacts on the Surrounding Population* (University Park: Pennsylvania State University Press, 1988), 2–5.

80. Delli Carpini and Keeter, *What Americans Know about Politics,* 323. The survey was no fluke: a nearly identical question asked the following year found that only 31 percent of the public knew that the answer was no.

81. "Beyond 'The China Syndrome,' " *Newsweek,* 16 April 1979, 31.

82. President's Commission on the Accident at Three Mile Island, *The Need for Change: The Legacy of TMI* (Washington, D.C.: Government Printing Office, 1979).

83. Mayer, *The Changing American Mind,* 106–108.

84. Rosenbaum, *Environmental Politics and Policy,* 246–59.

85. Energy Information Agency, *Annual Energy Review 1994* (Washington, D.C.: Government Printing Office, 1994), table 1.3.

86. Katz, *Congress and National Energy Policy,* 156.

87. Isser, 160–64.

88. G. Kevin Jones, "The Development of Outer Continental Shelf Oil and Gas

Resources," in *Energy Resources Development: Politics and Policies,* ed. Richard L. Ender and John Choon Kim (New York: Quorum, 1987), 97.

89. Gramling, *Oil on the Edge,* 121–26.

90. Riley E. Dunlap and Rik Scarce, "Trends: Environmental Problems and Protections," *Public Opinion Quarterly* 55 (1991): 651–72; Louis Harris, "A Call for Tougher—Not Weaker—Antipollution Laws," *Business Week,* 24 January 1983, 87; Paul R. Portney, "Natural Resources and the Environment," in *The Reagan Record,* ed. John L. Palmer and Isabel V. Sawhill (Cambridge, Mass.: Ballinger, 1984).

91. Morone and Woodhouse, *The Demise of Nuclear Energy?* 98–103.

92. Philip, *The Political Economy of International Oil,* 168–71.

93. Gramling, *Oil on the Edge,* 115–18.

94. J. Tyndall, "On Radiation through the Earth's Atmosphere," *Philosophical Magazine* 4 (1983): 200; Svante Arrhenius, "On the Influence of Carbolic Acid in the Air upon the Temperature of the Ground," *Philosophical Magazine* 41 (April 1896): 237–77.

95. William R. Cline, *The Economics of Global Warming* (Washington, D.C.: Institute for International Economics, 1992), 13–15; Lamont C. Hempel, *Environmental Governance: The Global Challenge* (Washington, D.C.: Island, 1996). Although the new attention given to global warming pleased many in the scientific community, they unanimously agreed that the hot summer of 1988 was *not* a sign of global warming.

96. Craig Trumbo, "Longitudinal Modeling of Public Issues: An Application of Agenda-Setting Processes to the Issue of Global Warming," *Journalism and Mass Communication Monographs* 152 (August 1995).

97. Ann Bostrom et al., "What Do People Know about Global Climate Change?" *Risk Analysis* 14 (1994): 959–69; Richard J. Bord, Robert E. O'Connor, and Ann Fisher, "In What Sense Does the Public Need to Understand Global Warming?" *Public Understanding of Science* 9 (2000): 205–18; Willett Kempton, "Public Understanding of Global Warming," *Society and Natural Resources* 4 (1991): 331–45; Willet Kempton, James S. Boster, and Jennifer A. Hartley, *Environmental Values and American Culture* (Cambridge, Mass.: MIT Press, 1995); Keith R. Stamm, Fiona Clark, and Paula R. Eblacas, "Mass Communication and Public Understanding of Environmental Problems: The Case of Global Warming," *Public Understanding of Science* 9 (2000): 219–37.

98. Bord et al., "In What Sense Does the Public Need to Understand Global Warming?"

99. Hempel, *Environmental Governance,* 104.

100. *Congressional Quarterly Almanac, 102nd Congress* (Washington, D.C.: Congressional Quarterly, 1993), 232.

101. Joanna Burger, *Oil Spills* (New Brunswick, N.J.: Rutgers University Press, 1997), chap. 4.

102. Delli Carpini and Keeter, *What Americans Know about Politics,* 75, 80.

103. Yergin, *The Prize,* 772–75.

104. T. M. Hawley, *Against the Fires of Hell: The Environmental Disaster of the Gulf War* (New York: Harcourt, Brace Jovanovich, 1992). See also Saul Bloom, John M. Miller, James Warner, and Philippa Winkler, *Hidden Casualties: Environmental, Health, and Political Consequences of the Persian Gulf War* (Berkeley, Calif.: North Atlantic, 1994).

105. *Congressional Quarterly Almanac, 102nd Congress,* 231, 250–58.

106. Patrick Lee, "Texaco Led Run-up in Southland Gas Prices," *Los Angeles Times,* 7 May 1996.

107. Ronald Brownstein, "Democrats Reassert Role of Government in Marketplace," *Los Angeles Times,* 2 May 1996.

108. David Leonhardt and Barbara Whitaker, "Higher Fuel Prices Do Little to Alter Motorists' Habits," *New York Times,* 10 October 2000.

109. "Crude Oozes Its Way Back to $30 a Barrel; Supply Boost Unlikely," *Los Angeles Times,* 13 May 2000, C1; Chris Kraul, "Oil Prices Jump Amid Renewed Mideast Troubles," *Los Angeles Times,* 13 October 2000, C1.

110. "Chevron Earnings Soar on Higher Oil, Gas Prices," *Santa Barbara News-Press,* 27 April 2000, A8; Ken Moritsugu, "U.S. Questions Refiners on Gas Prices," *Santa Barbara News-Press,* 13 June 2000, A5; "Clinton 'Worried' about High Gas Prices," *Los Angeles Times,* 17 June 2000, A22.

111. Pew Research Center for the People and the Press, "Rising Price of Gas Draws Most Public Interest in 2000," 25 December 2000 (http://www.people-press.org).

112. Frank Bruni, "Bush, in Energy Plan, Endorses New U.S. Drilling to Remedy Oil Prices," *New York Times,* 30 September 2000; Art Pine, "Climbing Gas Prices Expected to Be Hot Topic in Campaign," *Los Angeles Times,* 14 March 2000, A16.

113. Alison Mitchell, "Gore Says Bush Plan Will Cause Lasting Damage to the Environment," *New York Times,* 30 September 2000.

114. Frank H. Murkowski, "Let Alaskan Oil Help the State, Nation," *Los Angeles Times,* 17 February 2000, B9; George Reisman, "Government's Oil Prices," *Orange County Register,* 2 April 2000, 2.

115. Thomas S. Mulligan, "Tech Companies a Drain on Power Grid," *Los Angeles Times,* 12 December 2000, A1.

116. Peter G. Gosselin, "Most of the West in the Same Power Jam as California," *Los Angeles Times,* 26 February 2001, A1.

117. Peter Navarro, "Gas and Pipeline Costs Fueled Electricity Crisis," *Los Angeles Times,* 19 January 2001, B9; Chris Kraul, "Natural Gas Prices Rise as Worries over Energy Build," *Los Angeles Times,* 22 August 2000, A1; Marla Dickerson, "Manufacturers Reeling from Soaring Price of Natural Gas," *Los Angeles Times,* 17 December 2000, A1.

118. See the California Public Utilities Code and Bill No. AB1890, 1996, which can be found on the Web using the California State Legislature's search engine (http://www.leginfo.-ca.gov/bilinfo.html).

119. Nicholas Riccardi and Steve Berry, "Deregulation Didn't Foster Competition," *Los Angeles Times,* 7 February 2001, A1; Peter Schrag, "Blackout: Did California Just Mess It Up Badly, or Is Deregulation in Electricity an Inherently Bad Idea?" *The American Prospect,* 26 February 2001, 29–33.

120. Chris Kraul, "Antiquated Power Lines Add to Energy Woes," *Los Angeles Times,* 31 January 2001, A1.

121. Dan Morain and Mitchell Landsberg, "Davis Says State Will Buy Power for Resale to Utilities," *Los Angeles Times,* 14 January 2001, A1.

122. Christine Hanley, "If Northwest Gets the Chills, California Could Feel Weak," *Los Angeles Times,* 13 January 2001, A15.

123. Seema Mehta, "Fire-Damaged San Onofre Reactor May Not Be Back Online Till June," *Los Angeles Times,* 22 March 2001, A3.

124. Chris Kraul, "Power Crisis Generates Windfall for Suppliers," *Los Angeles Times,* 27 December 2000, A1; "Enron's Profit Surge Powers Energy Stocks," *Los Angeles Times,* 23

January 2001, C1; Greg Risling, "State's Wholesale Power Suppliers See Surge in Profits," *Santa Barbara News-Press*, 17 December 2000, A6.

125. Chris Kraul, "Charges of Gouging as Power Costs Skyrocket," *Los Angeles Times*, 8 August 2000, A1; Rick Lyman, "Power Shortage Fueling Cynicism," *Santa Barbara News-Press*, 13 January 2001, A1.

126. Mark Z. Barabak, "Most Californians Think Electricity Crunch Is Artificial," *Los Angeles Times*, 7 January 2001, A1.

127. Carl Ingram and Nancy Vogel, "Legislature OKs San Diego Electric Relief Package," *Los Angeles Times*, 31 August 2000, A1.

128. Nancy Vogel and Dan Morain, "Governor, Legislators Moving toward Bailout of Utilities," *Los Angeles Times*, 5 January 2001, A1.

129. Jennifer Kerr, "Voters Would Block Power Rate Hike, Davis Warns Wall Street," *Santa Barbara News-Press*, 3 March 2001, A3.

130. Morain, Dan, "Davis Walks a Political Tightrope on Energy Prices," *Los Angeles Times*, 12 September 2000, A1.

131. John O'Dell, "Power Crisis Is a Weapon in Electric Car Debate," *Los Angeles Times*, 19 January 2001, C1; Julie Tamaki and Jennifer Warren, "Groups Look for Silver Lining in Energy Crisis," *Los Angeles Times*, 18 March 2001.

132. Marla Cone and Gary Polakovic, "Bush's Idea of Easing Smog Rules Won't Help, Experts Say," *Los Angeles Times*, 25 January 2001, A18; Elizabeth Shogren, "Bush Drops Pledge to Curb Emissions," *Los Angeles Times*, 14 March 2001, A1; "Cheney Sees Benefits in Nuclear Plants," *Los Angeles Times*, 22 March 2001, A3; David Whitman, "The Coal Hard Facts," *U.S. News & World Report*, 26 March 2001, 17–18.

133. Nancy Vogel and Tim Reiterman, "PUC Approves Largest Electricity Rate Increase in State's History," *Los Angeles Times*, 28 March 2001, A1; Jenifer Warren and Miguel Bustillo, "With Energy Crisis Far from Over, Experts Say More Hikes Possible," *Los Angeles Times*, 28 March 2001, A1.

134. The Btu may be the measure most commonly used, but it is not perfect. Because people put different forms of energy to different uses, comparing all types of energy with a single measure is a bit crude. See William W. Hogan, "Patterns of Energy Use" in *Energy Conservation: Successes and Failures*, ed. John C. Sawhill and Richard Colton (Washington, D.C.: Brookings Institution, 1986), 30–31.

135. Richard Golob and Eric Brus, *The Almanac of Renewable Energy: The Complete Guide to Emerging Energy Technology* (New York: Henry Holt, 1993), 231.

136. Golob and Brus, *The Almanac of Renewable Energy*, 5.

137. For a fuller description of these various types of energy—especially renewable energies, see Golob and Brus, *The Almanac of Renewable Energy*.

138. Lee, Ball, and Tabors, *Energy Aftermath*, 43–48.

139. Rosenbaum, *Environmental Politics and Policy*, 260.

140. Mel Horwitch, "Coal: Constrained Abundance," in *Energy Future: Report of the Energy Project at the Harvard Business School*, ed. Robert Stobaugh and Daniel Yergin (New York: Random House, 1979), chap. 4.

141. Morone and Woodhouse, *The Demise of Nuclear Energy?* chap. 8.

142. Jeff Donn, "Nuclear Energy Running Out of Steam," *Santa Barbara News-Press*, 28 March 1999, A7.

143. Rosenbaum, *Environmental Politics and Policy*, 249.

144. Glenn Adams, "Federal-Ordered Destruction of Dam May Set a Precedent," *Santa Barbara News-Press*, 2 July 1999, A8; Kim Murphy, "Talk of Demolishing Dams Yields Torrents of Debate," *Los Angeles Times*, 21 June 1998; Kim Murphy, "U.S. Salmon Plan Could Lead to Removal of Dams," *Los Angeles Times*, 22 December 2000, A3.

145. Richman, "The Polls," 582.

146. Energy Information Administration, *Annual Energy Review 1997* (Washington, D.C.: Government Printing Office, 1997), xix.

147. Alan Tonelson and Beth A. Lizut, "If We Kicked the Oil Habit, Saddam Wouldn't Menace Us," *Washington Post*, 15 September 1996.

148. Colin J. Campbell and Jean H. Laherrère, "The End of Cheap Oil," *Scientific American*, March 1998, 83.

149. Tyler Marshall, "U.S. Dives into a Sea of Major Rewards—and Risks," *Los Angeles Times*, 23 February 1998.

150. William R. Cline, *The Economics of Global Warming* (Washington, D.C.: Institute for International Economics, 1992).

151. See Colin J. Campbell, *The Coming Oil Crisis* (Essex: Multi-Science, 1998), 67–68, from which this discussion of world oil reserve estimates is taken. See also Colin J. Campbell, *The Golden Century of Oil, 1950–2050* (Dordrecht, Neth.: Kluwer Academic, 1991).

152. Campbell, *Coming Oil Crisis*, 73.

153. Campbell, *Coming Oil Crisis*, 72.

154. *Oil and Gas Journal*, December 1995.

155. Campbell, *Coming Oil Crisis*; Craig Bond Hatfield, "Oil Back on the Global Agenda," *Geotimes*, 8 May 1998, 121; L. F. Ivanhoe, "Updated Hubbert Curves Analyze World Oil Supply," *World Oil* 217, no. 11 (November 1996): 91–94; C. D. Masters, D. H. Root, and E. D. Attanasi, *Science* 253 (1991): 146–52; John D. Edwards (University of Colorado, August 1997—cited by Campbell and Laherrere).

3

What Are the Trends
in Public Opinion?

Just as America's energy situation changed since World War II, public opinion about energy issues changed as well. Opinions, after all, are not held in isolation. Events occur, conditions change, and people respond. The growing attention to environmental problems, the growing dependence on foreign oil, and the wars, energy crises, and all the other events chronicled in chapter 2 influenced what people thought about energy policy. In this chapter, we follow those trends and begin to explain them.

Mapping out trends in public opinion is not just a matter of describing the past. If we can understand past trends, we can better predict the future. To put it in more concrete terms, as the world petroleum supply diminishes and tight supplies cause the price of oil to increase, will the public still oppose the development of more offshore oil and nuclear power? As we shall see, history suggests that the answers seem to be "Yes" to more oil but "No" to more nuclear power.

A NOTE ABOUT DATA

The record of what the public has thought about energy is not nearly as complete as the history of the events and policies related to energy. Before the 1973–1974 energy crisis, questions about oil, coal, gas, and nuclear energy were thought to be so dull and uninteresting that survey researchers almost completely ignored them. Indeed, before the mid-1960s, survey researchers ignored just about all environmental issues.[1] A few scattered surveys exist, but no systematic inquiries were conducted before the 1970s, and no repeated questions were asked that would allow us to learn how opinions changed over time. Our look at trends in public opinion toward energy issues, therefore, begins shortly after the beginning of the first energy crisis.

Using 1973 as the starting point for an examination of public opinion on energy issues is something of a problem, because it can mislead readers about opinion *before* 1973. The natural presumption might be that public opinion in previous years must have been similar to opinion in 1973. If support for energy development was high in the 1970s, readers might reason, it must have been high in the 1960s and 1950s as well.

That conclusion would be a serious mistake. Pollsters began asking questions about energy issues in 1973 because of the OPEC oil boycott and the energy crisis. Opinions about oil and other energy sources were hardly typical then. In a sense, this is precisely the wrong time to begin a trend analysis, because the starting point is so atypical. I do so because these data are the earliest available. Readers, however, should be wary of jumping to conclusions about public opinion in earlier decades.

Despite the lack of much hard data, we can draw some inferences about what the public thought on some energy issues before the first energy crisis. So, before turning to an examination of trends in public opinion since 1973, let me offer a few observations about what happened before the pollsters turned their attention to America's energy supply.

PUBLIC OPINION BEFORE 1973

The historical record presents evidence that environmentalist opinion existed long before Rachel Carson and the Santa Barbara oil spill helped launch the modern environmental movement. The record on offshore oil drilling is particularly clear. The critical editorials in the *Santa Barbara Daily Press* and the *San Jose Mercury News* quoted in chapter 2 show that at least some people opposed oil development along the coast as early as 1899. Whether the editorials represented widespread opinion is hard to say, but the Chamber of Commerce led efforts to end local oil drilling in Santa Barbara, and it is a safe bet that a good number of citizens joined them.

Our next indication of public opinion comes in the 1920s, when offshore oil development was the subject of a good deal of legislative struggle. Most of it focused on who would get the royalties from the drillers, but some people voiced objections to oil drilling itself. Finally in 1930, the public spoke clearly, voting a ban on new offshore oil drilling in state waters. That ban was upheld in two more votes in the 1930s, but it quietly ended as World War II began and the need for oil to power American armed forces became clear. It was not until the anti–oil drilling protests in 1966 and 1968 that another indicator of anti-oil opinion arose. These protests show that at least among some members of the environmental community, oil drilling was unpopular even before the 1969 Santa Barbara Channel oil spill.

The basis for opposing oil development may have been different at the turn of the last century than it is now. Turn-of-the-century objections to oil drilling, according to political scientist Robert Wilder, were focused "on the quality of life and on tourism."[2] One certainly hears those same concerns today, but one also hears other concerns as well—the potential for the oil industry to cause damage to the environment and to harm people. That is, safety concerns have joined aesthetic judgments in the litany of complaints against oil development along the California coast.

Yet whatever the basis for the dislike of offshore oil drilling, it is clear that since the beginning of the industry, some people have objected to it. At times, a great many people objected (certainly a majority in the 1930s). As chapter 2 argued, and contrary to the historical claims of some writers, the oil spill from Union Oil Company's Platform A in the Santa Barbara Channel did not suddenly create anti-oil sentiment. There had, after all, been anti-oil protests in the same area only a few years earlier. Instead, the spill pushed oil into the spotlight and made the dangers of drilling apparent to everyone who followed the nightly news. No doubt the spill also caused many people to turn against oil, but some anti-oil feeling had clearly been there all along.

The story of the public's response to nuclear power is quite different. Although a handful of environmentalists objected to nuclear power from the very beginning, there is little evidence of any widespread, organized opposition. Indeed, nuclear power drew a good deal of support from environmental organizations. As we have seen, the Sierra Club generally favored nuclear power as an alternative to fossil fuels and actually endorsed the 1961 proposal to build a nuclear power plant at Diablo Canyon, on the central California coast.[3] As historian Thomas Wellock puts it, "The public was not inclined to challenge without reason the new priesthood of engineers and physicists within the nuclear industry."[4]

Early survey research on the public's response to nuclear power is thin, but one 1956 Gallup poll showed that most people believed scientists' and the government's claims that nuclear power was safe. When asked, "Would you be afraid to have a [nuclear power] plant located in this community which was run by atomic energy?" 70 percent said "No," and only 20 percent said "Yes" (the other 10 percent were undecided).[5]

Throughout the 1960s and early 1970s, however, more and more environmental activists and organizations turned against nuclear power. The Union of Concerned Scientists formed in 1969 to oppose technological threats to the environment, especially from nuclear power. The Sierra Club began opposing more individual nuclear power plant proposals and in 1974 finally came out against all nuclear power. The 1973–1974 energy crisis probably had both advantages and disadvantages for nuclear power. On the one hand, the energy crisis increased demand for cheap energy. On the other hand, it focused the nation's attention on energy issues and created a situation in which the critics of nuclear power were more likely to be heard.

The anti–nuclear power advocacy organizations clearly played central roles in reducing support for nuclear power. From the early days of commercial nuclear power in the 1950s until the Three Mile Island accident in 1979, there were no major nuclear accidents, so far as the public knew.[6] Nuclear power plants had well-publicized reliability problems (temporary shutdowns), but no catastrophic accidents similar to the tanker and oil platform spills and fires that plagued the petroleum industry. Nevertheless, public confidence in the nuclear industry's safety clearly fell, and antinuclear sentiment grew. The only plausible explanation is that nuclear industry critics had persuaded the public that nuclear power was dangerous and undesirable.

In 1973, near the beginning of the first energy crisis, another polling organization asked the public what it thought about nuclear power plant safety and got a very different answer from the one given in 1956. The Roper organization asked, "There are differences in opinion about how safe atomic energy plants are. Some say they are completely safe, while others say they present dangers and hazards. How do you feel—that it would be safe to have an atomic energy plant someplace near here, or that it would present dangers?" Only 36 percent said it would be safe, while 41 percent said it would present dangers.[7] Because the 1956 and 1973 questions were not worded identically, we cannot conclude that the public's confidence was cut exactly in half. Yet given the growing criticism of nuclear power, we can certainly conclude that support for nuclear power began waning in the 1960s—before the first energy crisis.

Finally, although there are no survey data on people's attitudes about energy issues from the 1960s, some pollsters asked about other environmental issues. Those data show that the public already had pro-environmental sympathies— especially on clean air and clean water. For example, a national survey asked, "Would you like to see something done in your community or part of the country on . . . controlling air pollution?"; 81 percent of the public said "Yes." When asked about "controlling water pollution," support rose to 90 percent.[8] These and other results suggest that environmentalist opinion was widespread by the mid-1960s.[9]

We should be careful in generalizing too much from attitudes toward clean air, clean water, or other environmental issues to opinions on energy policies. Nevertheless, the public's general support for environmentalism is consistent with opposition to offshore oil drilling. Given the public's pro-environmental leanings, it would not be surprising to find that the 1969 Santa Barbara oil spill galvanized anti-oil sentiment. If nothing else, the spill certainly gave environmental organizations a boost by bringing another environmental issue to the nation's front pages.

In sum, we cannot say much for certain about public opinion before the 1970s because of the lack of public opinion polls; nevertheless, we can draw one important conclusion: The public was never in love with any energy source, save perhaps for nuclear power in its early days, before the construction of any commercial nuclear power plants. Other energy sources were apparently viewed less charitably. In particular, oil drilling went through several periods of intense public opposition long before the modern environmental movement. The support for energy development in the 1970s, therefore, may well have been a century-long high point in popularity for oil.

PUBLIC OPINION AFTER 1973:
THE DECLINING POPULARITY OF
ENERGY PRODUCTION

The energy crisis of 1973–1974 brought not only public attention to energy issues but that of pollsters as well. Ever since that first energy crisis, survey researchers have sought to find out what the public thinks about a range of issues, from gasoline prices to strip-mining coal. Their surveys included a series of questions about support for more oil drilling, more nuclear power plants, and more coal mining. Because the questions were asked repeatedly and wording of the questions did not change over time, we can use them to track trends in public opinion.[10]

Oil, nuclear power, and coal represent the three major policy options for increasing the domestic production of energy (which, of course, is why polling organizations have asked about them). Although scattered polls show that the public likes the idea of a variety of renewable and alternative fuels, few surveys have asked policy-related questions about alternative fuels (for example, "Should the government increase research funding?"), and none of those questions has been repeated often enough over time to allow long-term trends to be tracked. Similarly, there are no trend data available on major policy conservation issues, such as proposals for a carbon tax. As a result, I focus on oil, nuclear power, and to a lesser extent, coal.

NATIONAL TRENDS

In broad outline, public opinion toward energy production can be described very simply. During the 1970s—the years of the brownouts, the OPEC boycott, and the gas lines—the public looked quite favorably on both increased offshore oil production and nuclear power, and it was warming to the idea of strip-mining coal. The public's initial support for nuclear power, however, declined throughout the decade. After 1980, the peak of the second oil crisis, public support for all three types of energy production declined. Nuclear power continued its slide, and both offshore oil drilling and strip-mining coal became less and less popular as the years passed.

Figure 3.1 presents data from Cambridge Research International polls that summarize the American public's thinking about three major policy options for increasing domestic energy production.[11] The questions were: "I'm going to read you several proposals for dealing with the energy crisis, and I'd like you to tell me whether you generally favor or oppose each one.

- Expanding offshore drilling for oil and natural gas
- Building more nuclear power plants
- Increasing strip-mining for coal, even if it means damaging the environment."

Although the questions used for figure 3.1 offer a good basis for examining trends over time, they differ in a critical respect. The oil and nuclear power

Figure 3.1 Support for Oil, Nuclear Power, and Coal in the United States

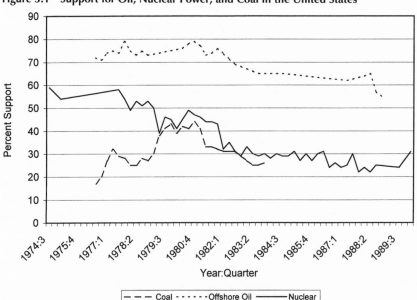

questions simply ask respondents whether they favor or oppose expanding those power sources. The coal question tells respondents that strip-mining may cause environmental damage. As a result, we need to be careful in comparing the results of the questions. We can safely compare the responses to any particular question over time, and we can compare the popularity of oil and nuclear power (because the questions have similar formats), but comparisons of people's opinions about coal with their opinions about oil and nuclear power are on shaky ground.

Some data suggest that coal may actually be a good deal more popular than nuclear power, contrary to the implication of the data in figure 3.1. A May 1979 Field Poll asked a representative sample of Californians a pair of questions that differed only in the choice of energy: "Suppose a power plant using [nuclear fuel/coal] was planned to be built in your county in order to supply electrical power in California? Would you favor or oppose its construction?"[12]

Fifty-four percent of the respondents said they would favor the coal-powered plant, but only 40 percent said they would favor the nuclear alternative. When the ideas of strip-mining and environmental damage were dropped from the question, coal came out as the preferred alternative. Of course, this survey was conducted in the immediate aftermath of Three Mile Island. Yet so was a CRI poll that showed coal and nuclear power were equally popular nationwide. Given the limited California data, it would be unwise to jump to the conclusion that coal was preferred either in the rest of the country or in any another year. Still, the California data do highlight the dangers of looking at the data in figure 3.1 and concluding that coal was the least-popular source of energy. That may well not be true.

Figure 3.1 shows the impact of various events on the public's thinking about energy policy. From 1976 though the end of 1977, public support for more offshore oil development rose. Although there were no dramatic events specifically related to oil during these years, this was the period during which President Jimmy Carter attempted to focus the nation's attention on energy policy. In the second two years of the Carter administration, support for more oil drilling ebbed slightly and then began to build as the second energy crisis developed. After the energy crisis and election of Ronald Reagan, support for offshore oil development slowly declined and continued to do so throughout the decade, falling from 79 percent in the last quarter of 1980 to only 55 percent at the end of the Reagan presidency in 1988. Still, at the decade's close, a majority of Americans favored more offshore oil drilling.

The overall trend in opinion toward nuclear power in the 1970s and 1980s was toward ever-lower levels of support, but the details are well worth a closer look. When CRI first asked about nuclear power in October 1974, 59 percent supported it—the highest level in CRI records. Support hit 58 percent in July

1977 but worked its way downward after that. By January 1979, only 50 percent of the public supported building more nuclear power plants. The March 1979 accident at Three Mile Island corresponds to a sharp drop in support for nuclear power. By April, only 39 percent thought more nuclear power was a good idea. But public support rebounded by July, rising to 46 percent—almost its January level. In the following months, support first declined but then rose as the second energy crisis began to develop. By July 1980, public support for nuclear power had returned to the pre–Three Mile Island level of 49 percent.[13] Yet even at the height of the second energy crisis, support for nuclear power fell short of the support it had received in 1974. The movement away from nuclear power was already evident by 1980.

As the crisis and the gas lines ended and as the price of oil fell in the following years, so did support for nuclear power. Indeed, the second large shift in this time series occurred between October 1981 and January 1982. Support for nuclear power fell from 43 percent to 32 percent—the same size as the post–Three Mile Island drop. Yet no dramatic events are associated with this second drop in support. Instead, support fell as the economy slid into recession after the 1981 peak in oil prices.

The next major nuclear accident occurred in the spring of 1986. In the April CRI poll, almost immediately before the meltdown of the Chernobyl nuclear power plant in the Soviet Union, support for more nuclear power in the United States stood at 31 percent. Three months later, it had dropped to 24 percent—a seven-percentage-point drop. Although the Chernobyl accident caused a huge number of deaths and an enormous amount of damage to the environment, while the Three Mile Island accident caused none, the Three Mile Island accident had a slightly larger impact on the American public. The lesser impact of Chernobyl is perhaps best explained by the journalistic rule of thumb, "Ten thousand deaths in Nepal equals a hundred deaths in Wales equals ten deaths in West Virginia equals one death next door."[14] To a great many Americans, at least, Three Mile Island seemed next door. Alternatively, the public may have listened to the nuclear-power experts insisting that the Chernobyl plant design differed enormously from the design of U.S. plants and that it could never happen here. In any event, these data show that although Chernobyl may have dashed the hopes of nuclear-industry advocates in the United States, it had only a minor impact on public opinion.

Finally, we need to take a step back and consider the role of the Three Mile Island and Chernobyl accidents on the overall decline in public support for nuclear power. Public support for nuclear power began to decline before Three Mile Island. It dropped right after the accident, but it quickly rebounded and within a little over a year, it had returned to its pre-accident level. Did the public shrug off the accident, or did some people decide to accept the risks because of high prices during the 1979–1980 energy crisis? The historical record alone cannot answer that question.

The Chernobyl accident corresponds with a further drop in support, but not a very large one. Moreover, the shift in opinion associated with Chernobyl could easily be lost among the many fluctuations in opinion during the 1980s. It certainly does not seem to have had a "devastating impact on public perceptions of nuclear risks," as one observer put it.[15] Together, the two accidents do not seem to explain the long-term decline in public support for nuclear power. They certainly must have contributed to it, but the decline began before the Three Mile Island accident. No complete explanation can ignore the role played by the antinuclear advocacy groups.

The last point to make about oil and nuclear power is that according to the data in figure 3.1, the public preferred increasing offshore oil development to building more nuclear power plants. Moreover, as the years went by, that ranking became more pronounced. In 1977, when CRI first asked about both offshore oil and nuclear power in the same poll, oil was 16 percent more popular. The gap grew over the years, as nuclear power lost support more quickly than oil. By 1988, the gap exceeded 40 percent.

The record of the public's attitude toward strip-mining coal is not as complete as the records for oil or nuclear power, but the survey data do cover the period when questions about coal mining were regularly in the news. During the Carter administration, more and more people decided that accepting some environmental damage in exchange for more energy was reasonable. As the second energy crisis developed, support for strip-mining rose, peaking at 44 percent in 1980. But as the energy crisis faded away and the Reagan administration took over, support for coal mining faded away as well, dropping to only 19 percent in 1985. That was the year of the third price shock, when oil prices collapsed and the American public largely abandoned its interest in energy issues. Although we have no survey data on coal after that year, it seems likely that strip-mining remained unpopular.

CALIFORNIA TRENDS

Opinion in California largely parallels national opinion. Energy development was generally popular in the 1970s. The 1979–1980 energy crisis saw a short surge in willingness to develop more energy sources, but that willingness did not last long. Ronald Reagan's entry into the White House corresponded with the beginning of a decade-long slide in support for oil development and nuclear power. The 1990 Persian Gulf War brought a reversal in the trend, but by the end of the 1990s both nuclear power and oil development were less popular than they had ever been.

Three measures of opinion are available for California—two about oil development, and one about nuclear power. Because California has no commercial coal mines, no questions were ever asked about coal mining. The Field Poll asked whether respondents agreed or disagreed with the following three statements:

- Oil companies should be allowed to drill more oil and gas wells in state tidelands along the California seacoast.
- Current government restrictions prohibiting the drilling of oil and gas wells on government parks and forest reserves should be relaxed.
- The building of more nuclear power plants should be allowed in California.[16]

When the Field Poll first asked about offshore oil development in 1977, a majority of Californians agreed that oil companies should be allowed to develop more oil along the coast, as figure 3.2 shows. Although Californians were clearly pro-oil, they were far less enthusiastic than other Americans; the CPI question found support to be about 20 percent higher nationally. We cannot be certain of the difference because the questions were worded differently, but given the similar content it seems reasonable to conclude that Californians were less

Figure 3.2   Californians' Support for Oil and Nuclear Power

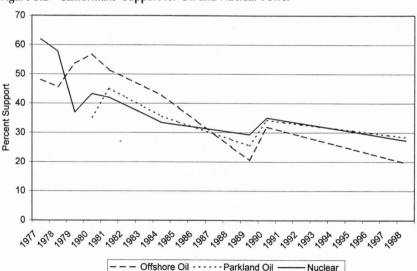

enthusiastic about oil drilling along their coast than were Americans living else-where.

The trend in attitudes toward offshore oil parallels the national trend. Sup-port for more oil drilling increased from 1978 to 1980 as the second energy crisis overtook the United States and then fell throughout the rest of the decade. The Persian Gulf War saw a resurgence of support for offshore oil, but in 1990 support peaked at 32 percent—a far cry from the 57 percent approval it had received in 1980. Finally, we should note that although there are no national data on attitudes toward offshore oil drilling after 1990, there are California data. They show that the popularity of offshore oil development continued to decline, reaching a dismal 20 percent by 1998.

The Field Poll also included a question about support for drilling in "govern-ment parks and forest reserves." With the exception of the beginning of the trend in 1980 and 1981, the pattern is similar to the offshore oil case. The prin-cipal difference is that over time, park and forest oil drilling went from being less popular than offshore drilling to more popular. The most likely explanation seems to be the politics of oil development. By the mid-1980s, environmentalists were loudly calling for a moratorium on new drilling along the California coast. President Reagan resisted the calls, but his successor issued an executive order banning new federal oil leases off the California coast. President Clinton contin-ued the ban, but legislation designed to make the ban permanent never passed Congress. The result was that the issue remained alive, with each new election seeing candidates, Democratic and Republican alike, swearing to uphold the moratorium. The continuing political attacks on offshore oil drilling, coupled with the almost complete silence over questions about drilling in parks and for-ests, seems to have moved public opinion against offshore drilling more sharply than against drilling inland.

Californians' attitudes toward nuclear power also follow the national trend, but the reaction to the 1979 Three Mile Island accident was sharper in the state than in the nation as a whole. The CRI polls showed an 11 percent drop in support for new nuclear power plants in the United States; the Field Poll showed a 21 percent drop from the previous year. Moreover, the early opinion trend in California was against nuclear power even before Three Mile Island. From 1977 to 1979, support for more plants fell from 62 to only 37 percent, somewhat more than the 19 percent drop registered nationally. Although nuclear power continued to decline in popularity in the 1990s, by 1998 it was actually more popular among Californians than offshore oil drilling.

A BROADER CONTEXT

To put the trend data in context, figure 3.3 presents two additional measures of environmental attitudes and a measure of the national mood, along with the

offshore oil and nuclear power data. The environmental measures are the percentages of Americans who say that they "would be prepared to sacrifice environmental quality for economic growth" and who say that we are spending too
much money "improving and protecting the environment."[17] The national-
mood measure, developed by James Stimson, is a composite index of trends in
a huge number of public opinion questions asked repeatedly over time. The
index is intended to reflect overall trends in opinion, or what Stimson calls the
national "mood." The scale of the index is arbitrary, and in figure 3.3 it has
been adjusted so that its average value is fifty and higher scores (as in all five
trends lines in this figure) indicate a more conservative, pro-development
mood.[18]

Four of the five measures in figure 3.3 follow roughly the same pattern: They
rise in the 1970s, reach a peak in 1980, and then fall throughout the 1980s.
Attitudes toward offshore oil development, environment-versus-economy trade-
offs, and environmental spending all moved roughly in tandem with the
national mood. That is, they all became more conservative and pro-development
in the 1970s and more liberal and pro-environment in the 1980s. These attitudes all roughly follow energy prices. The peak of support for more energy
development (1980) roughly matches the peak of energy prices (1981), and both

Figure 3.3   Selected Environmental Attitudes

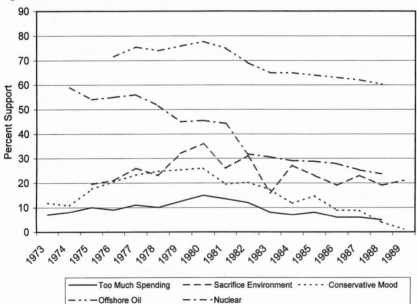

energy prices and support fell in the 1980s. The correspondence is not perfect, but it is close enough to warrant further investigation.

The exception involves attitudes toward nuclear power. Throughout both the 1970s and 1980s, public support for nuclear power fell. So while it seems that attitudes toward oil and coal moved with the broader tides, attitudes toward nuclear power had their own, independent causes. We shall explore those causes, along with the causes of attitudes toward oil and coal development, in the next section.

EXPLAINING THE TRENDS

Observing opinion trends is one thing; explaining them is another. A great many previous studies have described trends in opinion about environmental and energy issues. Most of these studies have pointed to historical events that seemed to explain changes in those trends. The OPEC oil boycotts caused public support for offshore oil drilling to rise; the Chernobyl accident caused public support for nuclear power to fall. This is the sort of analysis I presented in the first part of this chapter. While these arguments can be very persuasive, they are neither systematic nor rigorous. An investigator can, after all, mismatch events and public reactions. I am not trying to impugn any previous work here. The explanations of changing opinions that previous studies have proposed seem quite reasonable and well justified. Nevertheless, we can learn more about what caused changes in people's policy preferences if we use quantitative methods.

In the remainder of this chapter, I discuss and test formal models of changing opinion on energy issues. I begin with some well-known arguments about the relationship between age and environmental attitudes.

LIFE CYCLES AND GENERATIONS

Two of the most popular explanations of changing public opinion turn on the relationship between age and opinion. Whenever the young differ from the old—most commonly, whenever young people have more liberal opinions than their elders—someone suggests that when they grow up, the young will gain a better understanding of the issues and come around to their parents' point of view. An old maxim sums up the theory nicely: "If you aren't a liberal before you're 30 you have no heart, and if you are not a conservative before you're 40 you have no brain."[19] This is the *life-cycle* explanation—that aging causes people's opinions to change. An alternative explanation makes the opposite assumption about the effects of aging on opinions. According to some observers, people's opinions generally do *not* change as they get older. Consequently, as older generations die off and are replaced by younger people, public opinion as a whole comes to resemble the views of its successive younger generations. This is the *generational* explanation.

The life-cycle and generational models of change are relevant here because the young generally hold stronger, pro-environmental views than their elders. Jones and Dunlap, for example, find age to be "the best predictor of environmental concern—judged in terms of magnitude as well as consistency" in their examination of attitudes from 1973 to 1990.[20] The age gap raises two possibilities. If the life-cycle model were correct, aging would have little to do with changes in overall public opinion. That is, if the young start off as environmentalists but as they grow older come to resemble their elders, the result would be no net change in opinion over time. If the generational model were correct, however, a good deal would change. Older, pro-development generations would die off and be replaced by younger pro-environment generations, causing society as a whole to become more strongly environmentalist. This would help explain the past trend toward greater environmentalism and give us a basis for predicting that environmentalism will continue to spread for the foreseeable future.

The Life-Cycle Explanation

The life-cycle argument, that people become more conservative as they age, may sound attractive, but the evidence for it has always been thin. Children and teenagers of the 1960s who were brought up believing in racial equality did not, after all, become racial bigots as they grew older. The social scientists who considered the model in the 1950s and 1960s, however, regarded life-cycle explanations of attitude change as a serious possibility, because they lacked the data essential to test the model—repeated public opinion surveys asking identical questions over many years.

The best assessment of life-cycle effects arrived only recently, from William Mayer in *The Changing American Mind*.[21] Wanting to examine how public opinion changed over time, Mayer gathered every national poll he could find with questions that had been asked several times with identical wording over a period of years. The time-series data allowed him to search for life-cycle effects on a wide range of topics. Such effects, Mayer argued, should have at least two characteristics—the opinions must be correlated with age, and the opinions of the youngest generations must consistently move in the direction of the older generations.[22] For example, if we wanted to search for a life-cycle pattern in attitudes toward nuclear power, we would look first at surveys to see whether nuclear power was more popular among the old than among the young (or vice versa), and then we would follow a particular generation from one survey to another over time to see whether young people changed from opposing nuclear power to supporting it (or the opposite) as they grew older.

In his search for explanations of opinion change, Mayer found only three sets of questions with life-cycle effects—attitudes toward the income tax, welfare, and premarital sex. In all three cases, Mayer pointed to obvious reasons why people's opinions might change over the courses of their lives. In early adulthood, people's incomes and income taxes are fairly low. As their incomes and taxes grow, their dislike of taxes also grows. The young and the old hold more favorable views of welfare than the middle-aged presumably because the young and the old are more likely to receive welfare assistance themselves, while those in between are more likely to pay for it. Finally, attitudes toward premarital sex are undoubtedly affected by whether one thinks of oneself or one's children as the prospects for sexual activity. Not surprisingly, college-age people take a much more favorable view of premarital sex than those in their forties and fifties who have their own teenage children. Yet when college-age people grow older and have their own children, their views come to resemble those of their parents when they were middle-aged. In sum, we find life-cycle effects where there are changes over people's lives having obvious, direct connections to issues. We do not have to turn to vague arguments about people developing brains as they age (as the conservative maxim has it) or becoming more rigid and inflexible (as liberal critics suggest). According to Mayer, these obvious age-attitude connections are rare.

We now turn to a search for life-cycle changes in attitudes toward energy policy. As Mayer explains and we have noted, for the life-cycle explanation to hold, two patterns must appear in public opinion survey data. First, because the model claims that aging affects opinions, there must be a strong, consistent relationship between age and opinion. In the case of most environmental issues, this is true: The young are generally more pro-environmental than the old. Second, individual cohorts—people born in a particular span of years—must change over time in the direction of their elders. That is, if we use a series of public opinion surveys conducted once every ten years to follow the opinions of everyone born between, say, 1950 and 1959, we should find that they abandon their youthful environmentalism as they grow older. The data below in table 3.1 allow us to look for both those patterns in attitudes toward energy production among Californians.

The top panel in table 3.1 shows attitudes toward offshore oil drilling. The sample is divided into nine *cohorts*, or age groupings, which are identified in the leftmost column. The first cohort consists of people who were born between 1973 and 1980; the next comprises people who were born between 1964 and 1972; and so forth. The second column indicates how old these people were at the time of the first survey in 1981: The first cohort was from one to eight years

Table 3.1 Energy Opinions over the Life Cycle

| | | Pro–Offshore Oil Drilling | | |
| | | Percent Pro-Oil in | | |
Birth Years	Age in 1981	1981	1990	1998
1973–1980	1–8			22%
1964–1972	9–17		30%	17
1955–1963	18–26	42%	26	18
1946–1954	27–35	41	28	17
1937–1945	36–44	52	36	40
1928–1936	45–53	58	42	23
1919–1927	54–62	64	46	27
1910–1918	63–71	63	47	10
1850–1909	72–97	69	33	

| | | Pro–Parklands Oil Drilling | | |
| | | Percent Pro-Oil in | | |
Birth Years	Age in 1981	1981	1990	1998
1973–1980	1–8			39%
1964–1972	9–17		36%	24
1955–1963	18–26	40%	25	23
1946–1954	27–35	34	29	24
1937–1945	36–44	44	40	36
1928–1936	45–53	48	37	28
1919–1927	54–62	58	55	38
1910–1918	63–71	54	69	22
1850–1909	72–97	61	42	

| | | Pro–Nuclear Power Expansion | | |
| | | Percent Pro-Nuclear in | | |
Birth Years	Age in 1981	1981	1990	1998
1973–1980	1–8			26%
1964–1972	9–17		20%	
1955–1953	18–26	30%	32	23
1946–1954	27–35	33	38	31
1937–1945	36–44	42	36	19
1928–1936	45–53	55	47	37
1919–1927	54–62	50	47	21
1910–1918	63–71	56	65	9
1850–1909	72–97	53	50	

old, the second from nine to seventeen, the third from eighteen to twenty-six. This third cohort is the first one to be included in the 1981 survey, because the survey naturally included only voting-age respondents. The percentage of pro-drilling respondents in 1981, 1990, and 1998 are shown in the three columns on the right.

Reading down the 1981 column, we see that support for offshore oil drilling increases with age. Whereas 42 percent of the youngest cohort favored more drilling, 69 percent of the oldest cohort was pro-oil. Reading across the rows, however, we see that the young do not abandon their youthful environmentalism as they grow older. In 1981, 42 percent of those born from 1955 to 1963 supported more drilling. By 1990, support for more drilling had fallen to only 26 percent, and by 1998 it had fallen even farther, to only 18 percent. This pattern is repeated in all the age cohorts. Every age group becomes less supportive of offshore oil drilling as it ages from 1981 to 1998.[23] Instead of a life cycle, we see the opposite in these data—people becoming more pro-environmental as time passes. Contrary to conventional wisdom, as they grow older, the young do not become more like their parents—they become less like them.

The data on opinions on drilling for oil in public parklands and forest reserves, and on nuclear power (in the middle and lower panels of table 3.1) follow the same pattern. Older people are stronger supporters of both types of energy development than younger generations, but as each age cohort advances in years, its opinions shift in an anti–oil drilling, anti–nuclear power direction. In fact, between 1981 and 1998, opinions in every single age cohort in table 3.1 become less supportive of energy development.

The evidence presented here clearly points to the conclusion that the life-cycle model does not explain changes in attitudes on energy policy. Considering the general trend away from public support for more energy development, both in California and nationwide, this finding is hardly surprising. Other forces are clearly at work.

More generally, the life-cycle explanation seems to be a candidate for folk wisdom—a myth that persists despite scientific findings because it fills a social need. Older people who hold more conservative views no doubt feel frustrated by the fact that their children's generation disagrees with them. The folk wisdom that people grow more conservative as they grow older and wiser is surely comforting to them. It allows the older generation to say, "When you grow up, you'll come around to our point of view." It may be comforting, but it is not true.

The Generational-Replacement Explanation

An alternative to the life-cycle model is the generational-replacement explanation, which rests on the assumption that people's opinions generally do not change over time. In this view, because of generational replacement—older cohorts dying off and younger cohorts entering the adult population—public opinion as a whole moves in the direction of the young.

In some areas, we know that the generational model explains change. As an example, consider the increase in the average level of education. Almost all Americans complete their formal education by the time they are in their mid-twenties. Early in this century, most people went no further than high school. Following World War II, there was a huge increase in the number of people going on to college. The Vietnam War years saw even more students continuing through college. At the same time these younger adults were receiving college education, older Americans with fewer years of school were dying off. In effect, well-educated young people were replacing poorly educated old people, causing the average level of education to rise.

Our question about environmental attitudes, then, is whether they follow the same pattern as education. On the surface, the generational explanation is at least consistent with the trend in attitudes on energy issues. The young oppose oil drilling and nuclear power more than do the old, which matches the trend of growing opposition in the entire population. Are people's environmental attitudes fairly stable over time? Is some of the growth of environmentalist views explained by the fact that each younger generation is more pro-environmental than the previous one?

To test the generational model, we examine a series of public opinion surveys over time. In this case, we look to see how opinion would change if there were no generational replacement over time. The trick to doing this is to take a pair of polls (conducted years apart) and statistically weigh the results of the second poll so that it reflects the population of the first. The data in table 3.2 do just that for attitudes toward offshore oil development.[24]

The first two columns of table 3.2 show our age cohorts again. The third column shows the percentage of the 1981 population in each cohort. The first entry, for example, shows that 24 percent of the population was born from 1955 to 1963. The next column shows the opinions of those people in 1990. Of course, by 1990 there was a new generation of young adults, but in order to see what opinion would have been, we count the opinions of the 1955–1963 cohort as being 24 percent of the 1990 population. By weighing opinions offered by the 1990 respondents to match the 1981 population, we see what opinion would have been in 1990 if the makeup of the population had not changed since 1981. We compare this to the opinion of the real 1990 population to see how much of the change is caused by generational replacement. The mechanics are simply a matter of multiplying (or weighing) the opinions of each 1990 cohort

Table 3.2   The Effects of Generational Replacement on Support for Offshore Oil Drilling

Birth Years	Age in 1981	Percent of 1981 Population	Opinions in 1990	Product
1955–1963	18–26	0.24	24.5	5.8
1946–1954	27–35	0.21	25.9	5.5
1937–1945	36–44	0.13	32.7	4.3
1928–1936	45–53	0.15	39.5	5.9
1919–1927	54–62	0.13	41.6	5.2
1910–1918	63–71	0.09	45.2	4.2
1850–1909	72 +	0.05	31.1	1.6
		1.00	31.9	32.5

Actual opinion				31.9
Estimated opinion w/o generational change				−32.5
Generational change				−0.6%

Birth Years	Age in 1990	Percent of 1990 Population	Opinions in 1998	Product
1964–1972	18–26	0.16	16.7	2.7
1955–1963	27–35	0.18	17.6	3.2
1946–1954	36–44	0.20	16.6	3.3
1937–1945	45–53	0.15	40.1	6.2
1928–1936	54–62	0.11	22.9	2.5
1919–1927	63–71	0.11	26.7	2.9
1910–1918	72 +	0.08	9.8	0.8
		1.00	19.6	21.7

Actual opinion				19.6
Estimated opinion w/o generational change				−21.7
Generational change				−2.1%

by their 1981 population percentages and adding up the results. This is what the last column does.

The conclusion of our analysis in table 3.2 is that if there had been no generational replacement between 1981 and 1990, about 32 percent of the population would favor more offshore oil drilling—which is almost exactly the same as the actual level of support, 32 percent (the difference is lost in rounding error). To put it a different way, from 1981 to 1990 support for offshore oil drilling shrank from 51 percent to 32 percent—a 19 percent drop. Generational replacement accounted for a fraction of a percent (the rounding error) of that decline; other factors accounted for the additional 18 percent.

The lower panel of table 3.2 presents an analysis of change from 1990 to

1998. The results are essentially the same. Support for offshore oil development declined from 32 to 20 percent—a 12 percent decline. If there had been no generational replacement, support for more oil would have dropped to 22 percent anyway. Generational replacement accounted for about 2 percent, a trivial shift.

Similar analyses for oil drilling in government parklands and forest reserves, and for expanding nuclear power, show that generational replacement explains only a very small change toward the pro-environment side.[25] In the case of oil drilling in parks, support dropped from 45 percent in 1981 to 34 percent in 1990 and 28 percent in 1998. Generational replacement accounted for 1 percent of the shift from 1981 to 1990 and another 1 percent of the shift from 1990 to 1998. The net result is that generational replacement explained 2 percent of a 17 percent change in public opinion. In the case of nuclear power, support fell from 42 percent in 1981 to 35 percent in 1990 and to 27 percent in 1998. Generational replacement accounted for a net change of only 1 percent of the total decline.

To sum up: Generational replacement certainly affects public opinion, but the effect is quite modest—perhaps 1 percent per decade in attitudes toward energy development. Mayer found that in the case of some measures of environmental attitudes, generational replacement caused larger changes, but never enormous ones.[26] Generational change is a slow, incremental process. When attitudes swing by 20 or 25 percent in less than two decades, generational change can explain no more than a small portion of the change. For a complete explanation, we need to look elsewhere.

The Period-Effects Explanation

When social scientists examine changes in public opinion related to age, they normally describe the changes in terms of three causes—life-cycle changes, generational changes, and period effects. Life-cycle effects, we have seen, are theoretically caused by some aspect of the aging process itself; generational effects are caused by the different circumstances in which successive cohorts are socialized; *period effects* are everything that is left over. In more general terms, period effects are causes of opinion change that affect all cohorts at the same time and produce a general shift of public opinion in some direction.

If, as we have argued, the evidence on both life-cycle and generational effects clearly shows that these two models fail to explain the bulk of the change in opinion on energy issues from 1981 to 1998, then we must look for a period-effects explanation. The question is, what caused attitudes to change so rapidly? The general answers are obvious: gasoline prices and the economy. Let us attempt to find out exactly what it was about the economy that drove opinions.

MODELS OF CHANGING
OPINION OVER TIME

The fact that support for more oil and coal peaked at the height of the second energy crisis in 1980 points to the economy as a major cause of the changing level of support for energy development over time. When gasoline prices soared, offshore oil drilling and strip-mining coal became popular. When gasoline prices fell, so did support for both oil and coal. But was the price of gasoline the driving force behind the demand for more energy, or was it general inflation, or some other aspect of the economy? Further, if the economy influenced attitudes toward oil and coal, did it also influence opinions about nuclear power?

The next three tables explore the effects of changing economic conditions on attitudes toward energy sources by presenting the results of a set of time-series regression models estimated with the aggregate data from figure 3.1. (Readers not familiar with regression methods should see the statistical methods appendix, at the end of the book.) The dependent variable in each of the models is the percentage of people who supported oil, nuclear power, or coal strip-mining each year from 1973 to 1990.[27] The independent variables causing support in each equation are, first, a measure of the economy and, second, the "lagged" value of the dependent variable—that is, the value of the dependent variable from the previous year.

The theory behind the models is that the level of support for, say, offshore oil drilling will remain constant unless a change in the economy causes the level of support to rise or fall. That is, last year's level of support for oil drilling predicts this year's level. If the economy is good, support for oil development should fall; if the economy is bad, people will want cheaper energy, and support for oil development should rise.

Table 3.3 presents the time-series models for offshore oil drilling from 1977 through 1988.[28] In the first equation, two variables are used to predict each year's support for oil drilling—the previous year's support and percent change in the consumer price index (CPI). The key number here is the CPI coefficient. An increase of 1 percent in inflation causes a 0.77 percent increase in support for offshore oil drilling. Because inflation ranged from 2 to 13 percent per year during this period, this is a potentially substantial impact. The asterisk indicates that this coefficient does not quite reach the conventional level of statistical significance ($p < .05$), but it misses by only a hair ($p < .0502$). The adjusted R-square of 0.89 also indicates that the model does a very good job of explaining support for oil.[29] Realistically, inflation seems a likely cause of attitudes toward oil development.

The next equation in table 3.3 uses percent change in gasoline prices instead of inflation. The results are essentially the same. The coefficient of 0.12 means

Table 3.3   Time-Series Regression Models of Support for Offshore Oil Drilling

	(1) b	(2) b	(3) b	(4) b
Intercept	22.45	8.36	1.61	−7.93
Percent Pro-oil in previous year	.59**	.87**	.96	1.08**
Percent change in Consumer Price Index	.77*			
Percent change in gas price		.12*		
Percent change in real median income			−.11	
Percent change in real GDP				.46
Adjusted R²	.89	.88	.82	.85
N	12	12	12	12
Durbin's t	1.74	1.07	1.24	.46

Note: Durbin's t is a measure of autocorrelation. None of the Durbin's t statistics are statistically significant, indicating that autocorrelation is not a problem.
*p < .06; **p < .05

that a 1 percent increase in gasoline-pump prices causes support for offshore oil drilling to increase by about one-tenth of a percent. This coefficient may seem small, but gas prices rose and fell by as much as 25 percent during these years. Again, the coefficient just misses the conventional significance level (p < .056), and the high adjusted R-square indicates that the model performs very well. As a result, we cannot say which model is better. Given that gasoline prices and inflation are highly correlated with one another (r = 0.70), this is hardly surprising.

The models in the third and fourth columns of table 3.3 test two more measures of the economy—percent change in real median income per capita and percent change in real gross domestic product. The first measure is an inflation-adjusted measure of how people are faring in the economy; the second focuses on overall economic health, again inflation adjusted. Their coefficients are not even close to being statistically significant, and the R-squares are lower than in the inflation and gas-price models. In sum, these models show that inflation and gas prices—not the overall health of the economy—drive people's attitudes toward oil development.

Table 3.4 presents a similar set of time-series models for attitudes toward strip-mining coal. In the first equation, the percent of change in the consumer price index has a substantial influence on support for coal mining, but the previous year's support has no effect. The explanation for this finding seems to be that support for strip-mining coal was the product of the energy crisis and not much else. When energy prices drove inflation up, support for coal rose sharply. President Carter's focus on coal as an alternative fuel doubtless played an important part in this process. When inflation began falling in 1981, President Reagan

Table 3.4   Regression Models of Support for Strip-Mining Coal

	(1) b	(2) b	(3) b	(4) b
Intercept	14.21**	18.01**	30.61**	17.31
Percent pro-coal in previous year	.09	.39**	−.05	.48
Percent change in Consumer Price Index	1.76**			
Percent change in gas price		.32**		
Percent change in real median income			−2.25**	
Percent change in real GDP				−.77
Adjusted R²	.91	.83	.73	.47
N	10	10	10	10
Durbin's t	−1.00	−.41	.73	.98

Note: Durbin's t is a measure of autocorrelation. None of the Durbin's t statistics are statistically significant, indicating that autocorrelation is not a problem.
**$p < .05$

had entered the White House and the nation's focus on energy policy rapidly faded. Support for strip-mining faded along with it.

In the second equation in table 3.4, we see that change in gasoline prices does not work quite as well in predicting support for coal as inflation does. Previous year's support does have an effect here, and the adjusted R-square of 0.83 shows that this equation does not explain as well as the first equation.

The third equation in this set shows that percent change in real median income also influences attitudes toward coal mining. The coefficient is negative, meaning that for every 1 percent increase in median incomes, support for strip-mining coal falls by 2.25 percent. In simpler terms, as the economy improves, more people take the pro-environmental view. Although the coefficient makes sense, this model does not explain support for coal nearly as well as the inflation model does; the R-square is only 0.73.

The final equation performs even more poorly. The coefficient for change in gross domestic product is in the correct direction—that is, the negative sign shows that as the economy improves, support for oil declines. Nevertheless, none of the coefficients is statistically significant, and the explained variance is quite low, r = 0.47. Summing up Americans' attitudes toward strip-mining, we see that as with offshore oil development, the best predictor is inflation.

We now turn to nuclear power in table 3.5. We will examine the same set of models plus an additional model to consider the effects of the Three Mile Island accident. The pattern we find in the first four equations in this table does not match the pattern found with oil and coal. None of the economic measures has a statistically significant coefficient. Inflation, change in gasoline prices, change in median incomes, and change in gross domestic product all fail to achieve

Table 3.5   Time-Series Regression Models of Support for Building Nuclear Power Plants

	(1) b	(2) b	(3) b	(4) b	(5) b
Intercept	3.11	1.13	.12	1.06	25.24**
Percent pro-nuclear in previous year	.85**	.92**	.94**	.89**	.37**
Percent change in Consumer Price Index	.15				1.13**
Percent change in gas price		.04			
Percent change in real median income			.16		
Percent change in real GDP				.54	
Three Mile Island accident					−12.37**
Adjusted R^2	.90	.92	.92	.91	.95
N	16	16	16	16	16
Durbin's t	.20	−.27	−.40	.22	−.97

Note: Durbin's t is a measure of autocorrelation. None of the Durbin's t statistics are statistically significant, indicating that autocorrelation is not a problem.
**$p < .05$

significance. They are all too small. The only variable that works in the first four equations is the previous year's level of support, which predicts quite well.

The failure of the models is explained in the last equation, in which we add a dummy variable reflecting the Three Mile Island nuclear power plant accident. The variable is scored zero before 1979 and one in 1979 and afterward. Including it in the equation helps us answer the question, how did the public react to the Three Mile Island?

The public's reaction, according to equation 5, was quite negative. The coefficient of −12.37 indicates a 12 percent permanent loss of support for nuclear power. Moreover, once we take the influence of Three Mile Island into account, we find that inflation does affect support, just as it does in the cases of oil and coal. An increase of 1 percent in the consumer price index causes a 1.13 percent increase in support for nuclear power. Looking back at figure 3.1, we can see that two contending forces were operating at the peak of the second energy crisis in 1979 and 1980. Inflation was boosting support for energy up, while the memory of Three Mile Island was pushing support down. Once the energy crisis had passed, support fell steadily.[30]

Before moving on, we should also mention the 1986 Chernobyl nuclear plant accident in the Soviet Union. Unlike Three Mile Island, the Chernobyl accident did not seem to have any effect on U.S. public opinion. Several different models were estimated, but they all yielded the same finding—no effect. As our informal examination of figure 3.1 suggested, the shift in opinion following the Chernobyl accident did not stand out from other, normal fluctuations. The accident made the news, but the public did not care.

Let us pause here and try to put the findings in perspective. The models above all show that inflation and gasoline prices are the driving force behind support for more energy development. Inflation and gasoline prices are too closely related for our data to disentangle them, but there may be no point in doing so. The OPEC price hikes were a major cause of inflation during the 1970s, and people confronted inflation whenever they pumped gas.

The overall health of the economy, as measured by real median incomes and gross domestic product, seems to have had little effect on support for more energy development. Median income had some influence in predicting support for coal, but inflation worked better. Neither median income nor GDP worked in any other equation. To some extent, we can see this in figure 3.1. Support for more energy development peaked in 1980, when inflation peaked. Two years later, when the nation hit the bottom of a recession and unemployment exceeded 10 percent, support for energy development was already declining.

These findings suggest that people do not think of economy versus environment trade-offs in the way that some politicians present them. There certainly seems to be an environment-versus-inflation trade-off. Yet the connection between more energy development and lower energy prices is fairly direct. The supply-and-demand argument is an easy one to follow for many Americans. The broader claim that easing environmental regulations and opening up energy development are ways to improve the economy and help all Americans, however, does not seem to work. The Reagan administration made that sort of argument in the early 1980s, and the public responded with increasingly pro-environmental sentiments. Perhaps the argument that opening up more oil fields will help lower the unemployment rate is too much of a stretch. That type of economy-versus-environment trade-off argument does not seem to persuade the public.

CONCLUSION

The final question to address in this chapter is, what can past trends tell us about how the American public is likely to behave in the future? There are several useful answers.

First, the declining popularity of offshore oil development and nuclear power does not seem to be caused by some inexorable process, such as generational replacement. The young are indeed a good deal more pro-environment than older generations; nevertheless, the data indicate that generational replacement is a slow process that has had little influence on total public support for energy development. Consequently, the passage of time and the arrival of new generations will not necessarily push the public even farther into the anti-energy development camp. Environmentalists will not win simply by waiting long enough.

This conclusion does not necessarily generalize to other environmental issues. Mayer did find that generational replacement had some influence on attitudes toward environmental spending. Yet he also found that the principal causes of changing environmental attitudes were economic conditions, not generational replacement. The passage of time caused opinions to change a few percent; economic conditions, however, caused far larger swings in attitudes.[31]

Second, the price of energy—and more generally, the inflation rate—can cause huge shifts in support for all types of energy. If the United States again faces a sharp rise in the cost of oil or a broad run of inflation, we should expect to see support for all types of energy development increase.

That "if" is likely to happen. The experts predict, as discussed in chapter 2, that the world oil supply will largely be exhausted by the end of the century. More immediately, some experts suggest that the world will begin to see diminishing supplies and rising prices within the next ten to twenty years. Their line of reasoning is that the world will not suddenly run out of oil; rather, as oil fields become less productive and the process of recovering oil from more distant and dangerous places becomes more expensive, the price of oil will rise. That increase should trigger more support for domestic oil drilling—especially in convenient places like the coasts of California, Oregon, Florida, and the other southeastern states.

Back in the 1970s, a number of pollsters asked their respondents whether they thought that alternative sources of energy, such as wind and solar power, would largely replace oil and nuclear power within the next twenty years. Urged on by optimistic predictions from environmental scientists, many people said "Yes." Those hopes and predictions did not come true.

Since that time, the science of alternative energy has taken important steps forward. We are now able to produce wind and solar energy at competitive prices in some areas. In the next twenty years, however, if the optimistic predictions about alternative energy do not come true, the price of energy will begin a long-term increase, and the California coastline, along with many other currently off-limits areas, will begin to see oil rigs again.

Finally, we must consider the question of the future of nuclear power. Support for nuclear power also responds to inflation and the cost of energy, but it is far less popular than oil development. Moreover, its popularity does not seem to be recovering from the Three Mile Island accident. It will probably eventually recover, especially if oil prices rise substantially. Yet given that nuclear power has much more ground to make up than does oil drilling, there seems to be less hope for advocates of nuclear power than for advocates of oil. In fact, its best hope for a return to public favor is probably a combination of rising oil prices and the failure of alternative energy sources to produce enough power. Given a choice between nuclear power plants and wind, solar, or geothermal power, the public would no doubt prefer one of the latter alternatives. Given a choice between nuclear power and far higher energy prices, however, and the public might rethink its opposition to nuclear power. Still, on balance the outlook for the industry does not look bright.

NOTES

1. Riley E. Dunlap, "Public Opinion and Environmental Policy," in *Environmental Politics & Policy: Theory and Evidence*, ed. James P. Lester, 2d ed. (Durham, N.C.: Duke University Press, 1995), 70.

2. Robert Jay Wilder, *Listening to the Sea: The Politics of Improving Environmental Protection* (Pittsburgh, Pa.: University of Pittsburgh Press, 1998), 33.

3. Thomas Raymond Wellock, *Critical Masses: Opposition to Nuclear Power in California, 1958–1978* (Madison: University of Wisconsin Press, 1998), 68–72.

4. Wellock, *Critical Masses,* 22.

5. George H. Gallup, *The Gallup Poll: Public Opinion 1935–1971* (New York: Random House, 1972), 1400–1401. The safety question is the single relevant question in the only set of nuclear power questions Gallup asked before 1973.

6. We now know that there were major releases of radioactive gases near Hanford, Washington, in the 1950s, and that they caused substantial human harm—including birth defects and cancer. At the time, however, those accidents were covered up; no one, not even residents in the local communities, heard about them.

7. William G. Mayer, *The Changing American Mind: How and Why American Public Opinion Changed between 1960 and 1988* (Ann Arbor: University of Michigan Press, 1992), 491.

8. Mayer, *The Changing American Mind,* table 5.17, 486.

9. Mayer, *The Changing American Mind,* 102.

10. Because even small changes in the wording of public opinion questions can cause the results to change, to track opinion over time one must use identically worded questions.

11. Cambridge Reports/Research International (955 Massachusetts Ave., Cambridge, Massachusetts 02139) asked the three questions. The minimum sample size is nine hundred. The 1974–1988 data are from Mayer, *The Changing American Mind,* 350, 355, 489, 492. The 1989–1990 data on nuclear power are from Eugene A. Rosa and Riley E. Dunlap, "Nuclear Power: Three Decades of Public Opinion," *Public Opinion Quarterly* 58 (Summer 1994): 295–325.

12. These questions were asked by Field Institute in California Poll 7902. See the data appendix for details.

13. From the point of view of statistics, the 1 percent difference between the pre- and post–Three Mile Island figures is statistically trivial and insignificant.

14. Edwin Diamond, *The Tin Kazoo* (Cambridge, Mass.: MIT Press, 1975), 94.

15. Joseph G. Morone and Edward J. Woodhouse, *The Demise of Nuclear Energy? Lessons for Democratic Control of Technology* (New Haven, Conn.: Yale University Press, 1989), 100.

16. These questions were asked in Field Polls 7703, 7801, 7902, 8002, 8006, 8104, 8401, 8903, 9004, and in the California Offshore Oil Drilling and Energy Policy Survey. The sur-

veys were conducted by the Field Institute, a nonpartisan, not-for-profit public opinion research organization established by the Field Research Corporation (550 Kearny Street, Suite 900, San Francisco, California 94108). The samples were representative cross-sections of California adults, with sample sizes ranging from five hundred to 1,014. See the data appendix for details.

17. The first item is from Cambridge Reports International. The question is: "Which of these two statements is closer to your opinion? (a) We must be prepared to sacrifice environmental quality for economic growth, (b) We must sacrifice economic growth in order to preserve and protect the environment." The second question, from the NORC general social survey, is: "We are faced with many problems in this country, none of which can be solved easily or inexpensively. I'm going to name some of these problems and for each one I'd like you to tell me whether you think we're spending too much money on it, too little money, or about the right amount."

18. For a description and analysis of the index, see James A. Stimson, *Public Opinion in America* (Boulder, Colo.: Westview, 1991). The data are the "dyadic recursion" estimate of the mood, from appendix 2. For another interpretation of the index, see Eric R. A. N. Smith, "What Is Public Opinion?" *Critical Review* 10 (1996): 95–105.

19. Bob Egelko, "Conservative Justice's Opinions Raise Eyebrows," *Santa Barbara News-Press*, 19 October 1999, B4.

20. Robert Emmet Jones and Riley E. Dunlap, "The Social Bases of Environmental Concern: Have They Changed over Time?" *Rural Sociology* 57 (1992): 38.

21. Mayer, *The Changing American Mind,* chap. 3.

22. Mayer, *The Changing American Mind,* 178–79.

23. The exception here is the cohort of people born in 1909 or earlier. The 1998 survey did not include any of these people, who would have been at least eighty-nine by the time of the survey.

24. This method is used by Mayer, *The Changing American Mind,* chap. 7, and Paul R. Abramson, *Political Attitudes in America: Formation and Change* (San Francisco: W. H. Freeman, 1983), chap. 4.

25. These data are available on the author's Web page: http://www.polsci. ucsb.edu/faculty/smith/.

26. Mayer, *The Changing American Mind,* 176–78.

27. Because we do not have a complete set of quarterly data, I have chosen to use annual data rather than deal with missing-data problems. The results of quarterly and annual models are quite similar. None of the main conclusions would be changed with quarterly data. The annual data points are the average percentage support of all surveys for each year.

28. The data in figure 3.1 cover the years 1976 through 1988. The first year in this series, 1976, is lost to the model because there is no year before 1976 to use as the independent variable.

29. The Durbin's t at the bottom of the table is a measure of autocorrelation. The value of 1.74 is large, but not statistically significant, which conventionally means that the hypothesis of autocorrelation should be rejected. Because the coefficient is relatively large, however, I estimated models correcting for possible autocorrelation. The alternative models yielded almost identical coefficients and standard errors; therefore, I present only the original equation.

30. One might wonder whether the simple dummy variable distinguishing between pre-

and post–Three Mile Island (TMI) years is the best way to model opinion. So far as I can tell, it is. I estimated several alternative models, testing the hypotheses that the influence of TMI diminished over time (using Koyck and other distributed-lag models) and that the rate of decline in support for nuclear power changed after 1979. I also tested alternative models with median incomes and GDP. None of the alternative models performed nearly as well as the simple model in table 3.7.

30. Mayer, *The Changing American Mind,* 76–78.

How Much Does the Public Understand about Energy?

Public opinion is far more than just a matter of what the public as a whole thinks. Finding out what position a majority takes or which way opinion trends are moving is certainly a good place to start, but it is only part of the story. To understand fully what the public thinks about energy issues, we need to learn many more details. One of the keys to understanding what the public wants for U.S. energy policy is determining what the public knows about energy issues. Simply put, what facts do people know about this subject? Do people know enough to follow the Washington policy debates that newspapers cover? Do they know enough to assess the facts rationally and reason through various policy options? Are some of their factual beliefs wrong, and might these beliefs influence their opinions? Finally, if some misconceptions are widespread, what can be done to educate the public so that it better understands the facts and can make better policy judgments?

In this chapter, we will look at what the public knows about energy issues. We will begin by looking at Californians' knowledge of oil development and nuclear power.[1] We will then take a look at how knowledge differs in Santa Barbara and Ventura Counties—two coastal counties with offshore oil drilling, just north of Los Angeles. In these two counties, oil issues are regularly covered by the news media, and one might expect "not in my backyard" (NIMBY) attitudes to drive people to become relatively well informed about oil issues. Building on that foundation, we will investigate how people organize their opinions on energy issues—looking for evidence indicating whether people have thought enough about energy issues to assemble coherent views about energy policy. Finally, we will consider what might be done to raise the public's level of knowledge.

HOW MUCH DOES THE
PUBLIC UNDERSTAND?

One of the best-known findings of public opinion research is that most of the public does not pay much attention to politics and consequently does not know much about politics. The list of things not known by majorities is staggering. What do the words "liberal" and "conservative" mean? What majority in the House and Senate is needed to override a presidential veto? What are the first ten amendments to the U.S. Constitution called? Who is the chief justice of the U.S. Supreme Court? What are the names of the two U.S. senators representing your state? What was the Republican Party's "Contract with America"? What is the minimum wage? In nationwide, representative surveys of adults, less than half of the respondents were able to offer the right answers to any of these questions.[2]

We do not want to exaggerate these findings. My point is not that most people are ignorant. In fact, only a small portion of the population knows nearly nothing about politics. Rather, most people do know something about politics, though not a great deal. When Michael Delli Carpini and Scott Keeter used a huge number of information questions to construct a political-knowledge test for their sweeping study, *What Americans Know about Politics and Why It Matters,* they found that most people were somewhere in the middle. They were neither terribly ignorant nor terribly knowledgeable.[3] That is our starting point for understanding what the public understands about energy policy.

The implications with respect to knowledge about the U.S. energy situation and energy policy are bleak. There are a few aspects of the energy situation that should be easy to grasp for anyone—whether there are lines at gasoline stations, whether gas prices are rising, whether the United States imports oil from the Middle East. But most aspects of energy policy are complicated and difficult to understand without specialized knowledge. How much would new oil drilling along the California coast add to national and worldwide oil supplies? Can the United States ever become energy independent? What are the effects of gasoline price controls on the economy? Most recently, what is global warming, what causes it, and what are the policy options for doing something about it? None of these questions can be answered without a good deal of knowledge. In short, past studies suggest that if the energy situation is like other issues, we should expect the public to know something, but not a great deal, about it. Unfortunately, most energy policy issues are fairly complicated, and that is exactly the sort of knowledge that is not widespread.

The history of America's energy problems compounds the situation because it does not lend itself to a well-informed electorate. The energy crises of the 1970s certainly got a lot of attention, but they dominated the news for fairly brief periods of time. Before the energy shortages began in 1970, energy issues rarely moved beyond the business sections of newspapers. The fact that the United States depended on imported oil from the Middle East, for example, was rarely mentioned. The oil embargoes lasted for only a few months each, and the lines at gasoline stations did not even last that long. As a result, our energy problems received short bursts of news coverage, not sustained coverage, year after year.

Figure 4.1 offers a rough measure of the amount of news coverage given to energy issues from 1972 through 1995.[4] The figure shows the number of articles in the *Los Angeles Times* about petroleum and the petroleum industry. Most of these articles were in the business section of the paper, and many of them were no more than brief notes about mergers, promotions, new oil fields beginning production, and the like; nevertheless, they offered readers opportunities to learn about energy issues. The data, incidentally, are somewhat crude because they are the product of a publisher's indexing staff rather than a systematic content analysis by researchers.[5] Still, figure 4.1 illustrates the point that energy problems received bursts of attention during the oil crises of the 1970s but received less and less attention in the 1980s and 1990s. As previous research on news coverage of energy shows, when prices go up, so does news coverage. Price hikes—like bad news in general—get better coverage than price declines or good news.[6] That pattern comes out clearly in figure 4.1.

Our data begin in 1972, two years after the energy shortages and the long, hot summer of 1970 produced the first brownouts and began to set the stage for the first full-scale energy crisis. As energy shortages developed and prices rose in 1972, news coverage of petroleum issues began increasing. When the 1973 Arab-Israeli War broke out and the OPEC embargo began in October 1973, the number of stories skyrocketed, reaching 337 in December. But shortly afterward, newspaper interest in oil issues began to wane. By May 1974, the number of stories dropped under a hundred. By March 1975, as gasoline prices stabilized and the public's attention turned elsewhere, the number of stories declined again. Not until another Middle East war became imminent in 1979 did coverage of the subject surge upward again, but after peaking at over a hundred stories in June and November of 1979, the level of coverage fell once more. Between 1979 and the summer of 1990, there were rarely more than fifty items in a month—most of which were small items in the business section. The August 1990 Iraqi invasion of Kuwait sparked a brief interest in oil issues. During that month, the *Los Angeles Times* ran ninety-nine articles dealing with some aspect of the petroleum industry. But almost immediately after the Iraqi invasion—and long before the allied response and Persian Gulf War—newspapers began to ignore oil issues. Throughout the rest of the 1990s, petroleum received only scattered attention.

Figure 4.1   Petroleum Articles in the *Los Angeles Times*

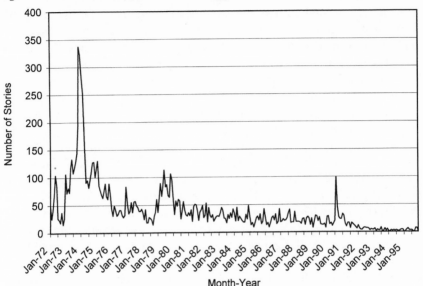

For a standard of comparison, think about the civil rights movement or the Vietnam War. These sorts of stories received sustained news coverage for years, not a few weeks or months. Repetition of stories on the same issues, often reviewing the facts for forgetful readers, helps people learn. Yet even with this sort of sustained coverage, large gaps in the public's knowledge appear. Consider, for instance, the news attention given to Newt Gingrich. Numerous stories about him appeared every day for weeks after the Republicans startled the nation by gaining House and Senate majorities in 1994. Even after the initial surge of attention subsided, Gingrich was in the news most days of every week through 1995 and 1996. Yet by May 1996, only half the public could supply his name when asked, "Who is the Speaker of the U.S. House of Representatives?"[7] Given examples like these, we can hardly expect energy policy, which received much less attention, to be well understood—especially the more complicated, technical aspects that are necessary for making informed choices.

When we turn our attention to the information questions about energy issues that have been asked of national samples, we find our expectations of low knowledge confirmed. Despite the Arab-Israeli wars, the gas lines, and the sharp gasoline price hikes, questions 1 and 2 in table 4.1 show that during the years of the energy crisis, only about half the public realized that the United States had to import oil to meet its energy needs. The failure to realize this most basic fact

Table 4.1   Selected Knowledge Questions—U.S. Samples

Question	Year: Percent Correct
1. Now, here is a list of some different kinds of materials and products. Would you go down that list, and for each one tell me whether you think we have adequate supplies of it in this country for our needs, or need to import some from other countries, or need to import most or all of it to meet our needs? . . .	1975: 54%
Oil?	1977: 49%
2. Do you think the United States has to import oil to meet our needs, or do we produce as much oil as we need?	1977: 48% 1979: 51% 1980: 63% 1991: 50%
3. Using the card, at the present time, how much of the oil used in the United States would you say we have to import from other countries? (Five choices.)	1977: 29% 1978: 30% 1981: 27%
4. What percentage of our oil do you think we now import? [Answers not read]	1977: 15% 1978: 12% 1990: 6% 1991: 5%
5. About what proportion of its oil would you say the United States imports from other countries? Would you say two-thirds or more, about half, about one-third, or less than one-third?	1984: 33% 1990: 49%
6. Here is a list of different companies. All of them have operations here in the United States. But, would you go down that list and for each one would you tell me whether, to your knowledge, the main headquarters is located here in the United States or located in a foreign country? . . .	
Exxon?	1978: 78% 1986: 76%
Shell?	1978: 17% 1986: 19%
7. From what you've heard or read, do you think a nuclear power plant accident could cause an atomic explosion with a mushroom-shaped cloud like the one at Hiroshima?	1979: 33%
8. Do you think that it is possible for a nuclear power plant to explode and cause a mushroom-shaped cloud like the one at Hiroshima, or don't you think that is possible?	1980: 31%
9. To your knowledge, what percentage of the nation's electric power is currently supplied by nuclear power plants?	1979: 5% 1986: 6%

Source: Michael X. Delli Carpini and Scott Keeter, *What Americans Know About Politics* (New Haven, Conn.: Yale University Press, 1996), app. 3.
Note: Delli Carpini and Keeter do not identify the surveys or sample sizes for the surveys above, but all surveys were conducted by major, reputable firms and are archived by the Roper Center for Public Opinion Research.

about U.S. energy problems during the 1970s may seem bizarre, but it is in line with other findings about the level of the public's knowledge generally. Large chunks of critical information are missing; relatively few people have much detailed knowledge about any public policy issue.

The other items in table 4.1 add to the picture of limited public knowledge about energy policy. Questions 3–5 offer different ways of asking how much oil the United States imports. In general, a relatively low percentage knows the right answer (although the survey firms conducting the polls scored "right" and "wrong" differently, so getting precise results is difficult). The 1990 result from question 5, that 49 percent got the right answer, seems surprisingly high considering that question 2 indicates that only about half the public even knows that we need to import oil—but we should keep in mind that questions 3–5 strongly imply that the answer is not zero (otherwise they would be trick questions, which is not what people expect from pollsters).

Questions 7–9 offer a glimpse of the public's knowledge of other energy issues. The two questions about nuclear power plant accidents are particularly interesting. The first question was asked shortly after the Three Mile Island accident (described in chapter 2); the second was asked a year later. Even though television, newspapers, and news magazines had reported extensively on what might happen during the Three Mile Island nuclear power plant accident, only a third of the public knew that a "mushroom-shaped cloud" was not a possible outcome. The Michael Douglas/Jane Fonda/Jack Lemon movie *The China Syndrome*, which appeared just before the Three Mile Island accident, may have been a critical and box office success, but an educational success it was not.

To focus more narrowly on California and to look at more recent data, we turn to the California Offshore Oil Drilling and Energy Policy Survey, conducted by the Field Institute in March 1998.[8] This survey asked a number of factual knowledge questions, shown in table 4.2. As with the national data, the results for California are anything but encouraging about the public's knowledge. We will begin by discussing the responses from the entire California sample and then turn to the Santa Barbara and Ventura samples.

The first question in table 4.2 asks how much oil is imported. The question is the same as one asked of national samples in 1984 and 1990 (question 5 in table 4.1), but the results are much worse. Only 26 percent of the California respondents knew the right answer—far lower than the 33 percent and 49 percent who got the right answer in 1984 and 1990. Although the California and U.S. samples are not the same and should be compared only with the greatest of care, it is hard not to speculate about the difference. The most reasonable explanation for the lower California number seems to be that knowledge has fallen all over the country, rather than that Californians are particularly uninformed. In 1984, the 1979–1980 oil crisis was still a relatively recent memory. In 1990, the Per-

Table 4.2   Selected Knowledge Questions—California Samples

Question	Percent Correct		
	California	Santa Barbara	Ventura
1. About what proportion of its oil would you say the United States imports from other countries? Would you say two-thirds or more, about half, about one-third, or less than one-third?	26%	29%	25%
2. As you know, the amount of oil in the world is limited. Do you know roughly how many years it will be before experts believe the world will run out of oil? [Probe if necessary]: Just your best estimate.	23	22	20
3. Do you happen to know whether it is currently safer to transport oil using oil tankers or using oil pipelines? By safer, I mean the way which is least likely to result in a major oil spill.	58	59	54
4. How often do you think a typical offshore oil platform along the California coast is likely to have a major oil spill—once every five years, once every ten years, once every twenty years, once every forty years, or less frequently than this?	15	14	16
5. When a major spill occurs, how much threat does it pose to human life—a great deal, some, only a little, or no risk at all?	6*	11*	7*
Sample N:	810	209	204

Source: California Offshore Oil Drilling and Energy Policy Survey, March 1998, conducted by the Field Institute.
*Statistically significant difference: $p < .05$

sian Gulf War was brewing, and the oil supply was once again receiving national news coverage. By 1998, however, oil issues were not in the news very often, and articles discussing how much oil the U.S. imports every year were rare indeed.

The second question in table 4.2 asks how long the world's oil supply will last. This is a tough question because the topic has not been the subject of many news reports since the 1970s; moreover, as discussed in chapter 2, the experts disagree. It has been the subject of some news coverage, however. Coincidentally, it was the subject of a *Scientific American* article released only a few weeks before the survey.[9] Despite this, half the respondents admitted they did not know the answer and declined to guess. The answers from the other half ranged from less than five years to over a thousand. (A reasonable estimate is somewhere between fifty and a hundred years—see chapter 2.) Taking any number in this range as correct, we find that only 23 percent of our sample knew the answer or could come up with a reasonable guess.

The next three questions in table 4.2 differ from the previous questions because they measure not only factual knowledge but the public's perceptions of risk as well. In the abstract, this difference may sound trivial, but in practice people who feel threatened respond by demanding that politicians do something about the risks. In the case of oil, those demands are to limit or end offshore oil development.

The first "risk" question asks respondents whether it is safer to transport oil by using oil tankers or pipelines. The answer, pipelines, should be fairly well known because most of the world's catastrophic oil spills have been tanker disasters—including such publicized ships as the *Castillo de Bellver,* the *Amoco Cadiz,* the *Torrey Canyon,* and the *Exxon Valdez.*[10] Because tanker spills have received so much news coverage, we should expect most of the public to know this answer, and they do. Fifty-eight percent of the California respondents correctly identified pipelines as the safest way to transport oil.[11]

The second risk question asks respondents to estimate how frequently a typical offshore oil platform will have a major oil spill. The public's answers to this question reveal seriously exaggerated fears. Twenty-three percent of the California public said that they expected a major oil spill every five years from a typical platform. An additional 25 percent said that they expected such a spill once every ten years. Although the exact meaning of "major oil spill" is not stated in the question, by any definition major oil spills from a typical platform are far less frequent than once every ten years. If we were to use a thousand barrels of oil as the standard for a major oil spill, we would have to conclude that the correct answer—given by only 15 percent of the sample—was "less than once every forty years," because there were only eleven spills of that size from all U.S. offshore platforms from 1964 to 1992 (and none since 1980).[12] Even using a far small number of barrels as the standard would still yield "less than once every forty years" as the correct answer, because of the huge number of offshore oil platforms in operation in U.S. waters (over 3,800 in 1990).[13] This exaggeration, or lack of knowledge, presumably contributes to the public's opposition to further offshore oil drilling.

The answers to the next question about oil spills also help us understand people's fears. The question asks how much threat a major oil spill poses to human life. Again we see that the public seriously exaggerates the threat posed by oil. Thirty-one percent of the statewide sample responded that a major oil spill would pose a serious threat to human health. An additional 36 percent believed that it would pose "some" threat. In fact, oil spills pose only a negligible threat to human life.[14] They certainly pose a substantial threat to many other kinds of life, but not to human beings. Indeed, if this were not true, the oil that endlessly washes up on the Santa Barbara shoreline from natural seeps in the Santa Barbara Channel would pose a major environmental hazard.

We turn now to the Santa Barbara and Ventura County data shown in the two right-hand columns of table 4.2. The chief observation to make about these data is that they are almost identical to the statewide figures. In fact, the percentage of correct answers to the first four questions—how much oil the United States imports, how long the world oil supply will last, whether tankers or pipelines are safer, and how often offshore oil platforms have spills—are statistically identical to the statewide percentages. The only question showing any statistically significant differences is the last one, about the health risk associated with oil spills—Santa Barbarans are more likely to know that oil spills do not threaten human life than are people living elsewhere in California, including Ventura County. Yet a difference of only 5 percent is substantively small. In fact, it is surprisingly small—especially given that a typical trip to the beach in Santa Barbara ends with cleaning the oil and tar off one's feet with paint thinner or baby oil (a fact that the tourist brochures somehow never seem to mention). So although Santa Barbarans know more about oil development than Californians living elsewhere, the difference appears only on one question, and the difference is small. There is certainly nothing here to suggest that Santa Barbarans living with oil development in their backyard know much more about it than anyone else.

ATTENTIVE PUBLICS

Let me take a step back and place these findings in the context of research about public opinion. Since the 1950s, scholars have known that a huge portion of the public is poorly informed. When the pollsters first discovered the public's low level of knowledge, the findings distressed most observers. They asked one another, "Can democracy survive with an uninformed public?" Eventually they focused on the facts that although the average citizen did not know much about political issues, a fair percentage of the public was well informed, and those who were knowledgeable were also more likely than the poorly informed to vote and become active in politics in other ways. Besides, democracy had survived long before pollsters began inquiring about how well the public understood politics.

Public opinion scholars also speculated that most people paid attention to some issues, but that the issues changed from one group to another, so that overall the public seemed poorly informed. Farmers, for instance, were thought to be knowledgeable about farm policies, auto workers about tariffs on Japanese and German cars, students about education policy, and so on. These groups of well-informed people were known as *attentive publics* or *issue publics*.[15] This idea comforted those who worried about democracy because it implied that the people having the greatest influence over different policies were proportionately well informed.

Although if the idea of attentive publics was comforting, it was also based as much on wishful thinking as on hard evidence. Belief in the existence of attentive publics came primarily from anecdotes rather than from public opinion surveys. After all, surveys asked everyone the same factual-knowledge questions, which did not leave many opportunities for investigating the existence of narrow attentive publics. Even though the evidence was weak, however, scholars generally accepted the existence of attentive publics until the 1980s, when interest in how much the public understood about politics renewed.

The 1980s saw several efforts to investigate the attentive-publics hypothesis. Two of the earliest and best known studies were based on the same survey—a 1985 pilot study leading up to the 1986 National Election Study. One of the principal investigators, John Zaller, concluded that political knowledge was a general trait.[16] People who knew a great deal about one subject, foreign policy for example, were likely to know a great deal about all other areas. Conversely, those who knew very little about any given area were likely to be uninformed in other areas as well. The other principal investigator of the 1985 NES Pilot Study, Shanto Iyengar, came to a slightly different answer.[17] Although he admitted the evidence was weak, he concluded that the data indicated at least some tendency for knowledge to be "subject-matter specific." That is, Iyengar argued that some people really did seem to specialize in different subjects.

More recently, Delli Carpini and Keeter's more thorough investigation of the public's knowledge of politics sided largely with Zaller. While they found some evidence for a slight tendency for people to learn more about specific areas, overall they concluded that describing political knowledge as a single trait was accurate. There were deviations showing, for example, greater knowledge of racial issues among blacks, but the deviations were minor. In general, Delli Carpini and Keeter found that people who knew a good deal about one subject knew a good deal about other political subjects as well, while those who were uninformed in one area were similarly uninformed about other areas also.

Despite the research questioning the existence of attentive publics, one could counter that researchers looked for attentive publics that were too broad, far broader than the original idea implied. Zaller and Iyengar examined whether some people focused on foreign policy, economic policy, or race-related policies—all of which are broad areas. Delli Carpini and Keeter's analysis was more detailed because they were able to work with surveys designed specifically to examine what people knew about politics. Still, they also asked about fairly broad topics. In their primary analysis based on their 1989 "Survey of Political Knowledge," they focused on questions that were grouped into the categories of "rules of the game," the substance of politics, political parties, people in politics, and gender issues.

The first four areas obviously take in broad ranges of political knowledge. "Gender issues" sounds sufficiently narrow and reasonably like the original idea of attentive publics, but the category actually consists of two questions—one about abortion and the other about when women won voting rights.[18] One can imagine a group of people who care deeply about the politics of reproductive rights, but appropriate questions for such a group would not include one about women's suffrage. Instead, one would expect that such a hypothetical attentive public would specialize in knowing such facts as when *Roe v. Wade* was decided by the Supreme Court, during what trimesters *Roe v. Wade* allows abortions, what *Webster v. Pennsylvania* decided, what a "bubble ordinance" is, what RU-486 is, and the names of groups that are among the principal supporters and opponents of abortion rights. Unfortunately, Delli Carpini and Keeter's survey did not include questions that would allow a search for this sort of narrow attentive public focus on a single policy.

Delli Carpini and Keeter also performed other analyses, but all of them are subject to the criticism that they treat attentive publics too broadly. In a reanalysis of the 1985 NES Pilot Study, they looked at foreign issues, economic issues, and race-related issues, as Zaller and Iyengar had done. In two smaller surveys of Virginia residents and of Richmond residents, they sought to discover whether some people focused on state or local issues, while others focused on national issues. To do so, they asked questions about national politics and about how state and local governments were run (for example, how the mayor is elected, how the city manager is elected). Although these latter questions were narrower, they still do not really match the initial idea behind attentive publics—that some people go out of their way to learn a great deal about a few narrow policy issues that interest them, but ignore politics in general. In sum, although Delli Carpini and Keeter's analysis went further than any preceding work, it left the door open for the possibility that groups of people form attentive publics that focus narrowly on specific issues of interest to them.

The best evidence in support of the existence of attentive publics is indirect. One way to conceptualize attentive publics, according to Jon Krosnick, is to think of them as groups of people who see particular issues as highly important.[19] People who think issues are especially important are likely to think about those issues often, to resist changing their opinions on those issues easily, to give great weight to those issues in choosing presidential candidates, and to learn more about those issues than others do. Krosnick's evidence clearly supports the existence of attentive publics on the first three of the four standards. Unfortunately, his surveys did not have knowledge questions for each of the issue areas he was studying. As a result, we have evidence supporting the existence of attentive publics (defined as people who assign special importance to particular issues) but no evidence indicating that they are especially knowledgeable. It would certainly seem to make sense that these attentive publics know more than other people, but the evidence to support that claim does not exist.

In sum, the evidence on attentive publics is mixed. Studies focusing on political knowledge have not found any clear evidence that attentive publics are especially knowledgeable in narrow areas—although arguably they were not looking in the correct way. Studies focusing on issue importance have found evidence supporting the existence of attentive publics, but they have not been able to measure how knowledgeable these people are.[20]

The evidence presented in table 4.2 addresses the question of the existence of attentive publics. The data cannot tell us whether they exist or not, but they can tell us that a group of people whom we might reasonably expect to be members of an attentive public and especially knowledgeable about the subject of offshore oil drilling are, in fact, not particularly knowledgeable. Let us put the findings in context by reviewing some history.

Santa Barbara has a hundred-year history of fighting with oil companies. The 1969 Santa Barbara Channel oil spill, about which local residents receive regular reminders, helped launch the modern environmental movement. Both local environmentalists and the county government have fought long, expensive battles against the offshore oil platforms, which are visible all along the coast, where most of the residents live. Environmentalists and the county government have also fought to prevent oil tankers from using the Santa Barbara Channel as a shortcut up and down the coast. The local newspapers regularly cover oil issues—offering their readers information on everything from gasoline prices to possible new oil developments, to the smallest oil spills. In 1994 and 1995 a major controversy erupted over a proposal by Mobil Oil Corporation to build an onshore facility designed to tap an offshore reservoir in state waters, using "extended reach" technology.[21] Both environmental groups and Mobil put on major public-relations efforts to win the battle. 1995 also saw anti-oil forces win a hotly contested, countywide ballot initiative to restrict further oil development in the county. Again, both sides poured volunteer time and a huge amount of money into their campaigns to educate and persuade the public. As a consequence of all these events, Santa Barbara looks like the perfect place to find a NIMBY-driven attentive public. Despite all this, Santa Barbarans know no more facts about oil politics than typical Californians. If we were to find evidence for an attentive public, this looks like the perfect test case. But the evidence is not there. No attentive public appears.

Although the evidence in table 4.2 shows that Santa Barbarans do not make up an attentive public, it remains possible that there are people scattered all over the state who take a special interest in the politics of offshore oil development.[22] That is, attentive publics may comprise people who find an issue especially important, irrespective of where they live, as Krosnick argues. To test this possibility, we need to focus on the issues themselves, rather than singling out Santa Barbara or Ventura and looking for greater knowledge there. Ideally, we should search for this hypothetical attentive public by using questions about the importance of offshore oil development. Unfortunately, we do not have any survey data about the importance of issues, so we must use the approach of Zaller, Iyengar, Delli Carpini, and Keeter.

Following this path, our first step was to perform an exploratory factor analysis of the knowledge questions from table 4.1 to see how many underlying factors cause both general knowledge and oil-related knowledge. Factor analysis is a statistical method that allows one to discover whether a set of observed variables, such as ability to answer correctly our information questions, is caused by one or more unobserved variables, or *factors*. If no attentive public exists, then only one factor should emerge—reflecting general knowledge of politics. If an attentive public does exist, two factors should emerge—one reflecting general knowledge of politics and a second reflecting specialized knowledge of oil issues. In addition, the numbers indicating the size of the influence of the factors on the specific questions, or *factor loadings*, should also be large.

Although this may sound a bit complicated, the underlying idea is simple. Factor analysis will tell us whether we need one or two "yardsticks" to describe how knowledgeable people are. If no attentive public exists for oil, then one yardstick will do. It will tell us whether people are well or poorly informed. If an attentive public exists, however, we will need two measures—one to tell us how knowledgeable people are about general political issues, and a second to tell us how knowledgeable they are about oil-related issues. If an attentive public exists, after all, then we should find some people who are poorly informed in general but know a lot about oil issues, and also some people who are generally well informed but know little about oil. It all comes down to how many factors best describe the data. (For a more detailed explanation of factor analysis, see the statistical methods appendix.)

Table 4.3 presents the results of two factor analyses of the five information items in table 4.2 plus five other questions that Delli Carpini and Keeter recommended as a general information test.[23] The first column presents the analysis of the California sample; the second column presents the analysis of the smaller Santa Barbara/Ventura sample. The five general information questions are listed first. As expected, the factor loadings for these five items, beginning with .65 for the question asking what job Al Gore holds, are fairly strong. That is, the unob-

Table 4.3   Factor Analysis of Information Items

	California Sample Factor Loading	Santa Barbara/Ventura Sample Factor Loading
What job does Al Gore hold?	.65	.66
Who decides if a law is constitutional?	.57	.53
What vote needed for veto override?	.52	.48
Which party has House majority?	.50	.55
Which party is more liberal/conservative?	.60	.63
What percent of our oil is imported?	−.08	−.08
How long will world oil supply last?	.24	.22
Tanker or pipelines safer to transport oil?	.15	.15
Frequency of oil platform spills?	−.09	.12
Danger of oil platform spills?	.16	.18
1st eigenvalue	2.61	2.65
2nd eigenvalue	.26	.32
N	810	413

Source: California Offshore Oil Drilling and Energy Policy Survey, March 1998, conducted by the Field Institute.

served variable "political knowledge" does a good job of explaining how well people do in answering these questions.

The five oil-related questions follow, with far weaker loadings. The weaker loadings indicate that the oil questions are poorer measures of general knowledge than the five questions recommended by Delli Carpini and Keeter. In fact, in the California sample, the two negative loadings indicate that the more people know, the *less* likely they are to know what portion of our oil is imported and how frequent oil spills from platforms are. These results reflect the facts that the oil questions are much tougher than the general information questions and that the public is less likely to know the answers. Despite the low knowledge of oil, however, the evidence points to the conclusion that a single underlying trait— political knowledge—explains all of the questions. A formal statistical test for the existence of a second factor rejected the hypothesis—that is, found that there was no second factor—in both the California and Santa Barbara/Ventura samples.[24] Moreover the small sizes of the second eigenvalues—far below 1.0—show that no real case can be made for two underlying factors in either of these data sets. In short, the data presented here support Zaller's and Delli Carpini and Keeter's claim that knowledge is a general trait and that attentive publics do not exist.

The second step in searching for people who specialize in knowledge about oil-related issues was to run a confirmatory factor analysis testing the possibility that a second unobserved variable might be contributing to people's ability to answer the questions. That is, rather than merely accepting the factor analysis at face value, we estimated several confirmatory factor-analysis models to see whether there is a general knowledge factor and a second, weaker oil-knowledge factor.[25] None of the models worked at all. In fact, as readers who themselves have worked with confirmatory factor analysis may have guessed from the low factor loadings of the oil items in table 4.3, not only did the models not explain anything extra, they fundamentally failed to work—in some cases producing implausible results. In sum, nothing in this analysis provides any evidence that a significant number of people are especially likely to know about oil issues.

We can sum up the findings on attentive publics quite easily: There is no evidence for their existence in any of the data we have examined. The oil questions in table 4.2 are certainly tough for the general public; nevertheless, if an attentive public focusing on oil politics exists, we should expect its members to know whether the United States actually needs to import oil, how much oil is left in the world, what the safest way to transport oil is, something about the frequency of major oil spills, and whether oil spills pose a threat to human beings. The concept of an attentive public, at least as most scholars understand it, does not depend on a localized, not-in-my-backyard response. Yet common sense tells us that if anything, NIMBY responses should make attentive publics especially likely to appear in places like Santa Barbara and Ventura counties. But none did.

One remaining possibility is that a well-informed, attentive public exists, but is quite small. Groups of only a few percent of the population would have escaped notice in my analysis. The number of political activists who focus on any particular issue is extremely small as a percentage of the entire population. If we were to take that number and double or triple it, it would still barely affect the results of any of our data analyses. As a result, we cannot make any claims about whether they exist or not. We can, however, suggest that well-informed attentive publics may be far smaller than previous investigators have suggested.

THE STRUCTURE OF OPINIONS

Our findings about the public's lack of knowledge about energy issues leads us to inquire about the consequences of low information for people's opinions on issues. If a sizeable number of people are so poorly informed, can they make sense of energy policy? Does the evidence suggest that people have carefully thought through energy production and conservation problems and developed coherent views on them, or that they have poorly considered opinions and inconsistent positions on the issues?

When we examine opinions on issues taken one at a time, we have no real basis for assessing the extent to which people's knowledge influences their opinions. If we were to look only at a question about support for offshore oil drilling, for example, we might imagine that the public holds a consistent set of preferences regarding energy policies and that their preferences about offshore oil development fit into a coherent strategy for dealing with energy problems. We might imagine it, but we would have no evidence of it. When we examine entire sets of issues at the same time, however, we can see evidence of the influence of knowledge on opinions—at least indirectly.

In this section, then, we will investigate how people fit together their opinions about energy issues. This analysis will give us a basis for drawing some conclusions about how much people understand about the issues, how they think about them, and how poorly policy makers and the public understand one another.

Conserving Your Energy and Using It at the Same Time

We begin our examination of how people's opinions fit together with a brief look at results of a public opinion poll conducted in California in 1989. These data are over ten years old, yet there is little reason to expect much substantive changes in people's opinions on these issues. The point of this exercise is to show possible inconsistencies in people's preferences. We will move on to a more systematic and conventional analysis in the following section, but these data merit discussion because they raise some interesting questions about the public's preferences.

These results, as we shall see, reveal that a significant number of Californians want to have their cheap gasoline and energy, but do not want to produce it—at least, not in California. These opinions do not contradict one another because the questions focus on energy production in California, rather than the entire United States. One can read them, therefore, as pointing to the extent to which people want to be free riders—having a benefit without bearing the costs. Still, these results may suggest to some readers that there really are people with contradictory opinions—those who want to have their oil and conserve it too, in a classic logical contradiction.

The first question in table 4.4 shows that in July 1989, 87 percent of Californians considered the country's energy situation to be extremely serious or somewhat serious. These results match those of a series of earlier surveys from 1979 through 1984, in which 89 percent or more of the Field Poll's respondents agreed with the statement: "Even with strict conservation, we will have to develop a lot more energy sources to meet this country's future energy needs." In short, the public believes an energy "crisis" exists and that something needs to be done about it.

Table 4.4   Selected Attitudes on Energy Issues

"How serious do you think this country's energy crisis is right now?"

Extremely serious	46%
Somewhat serious	41
Not too serious	9
Not at all serious	2
No opinion	2

"What about the offshore coastal areas of the United States. How much oil is there that could be discovered and developed in these areas?"

A great deal	27%
Quite a bit	20
Some	19
Only a little	9
No opinion	25

"How important do you believe it is for the United States to cut down on the amount of oil that it imports from foreign countries?"

Extremely important	48%
Somewhat important	37
Not too important	11
No opinion	3

"We can have oil drilling in offshore coastal areas and at the same time provide adequate environmental safeguards."

Agree	50%
Disagree	48
No opinion	2

Source: Data from the Field Institute's California Poll 8903 (July 1989).
Sample size = 993.

The second and third questions in table 4.4 suggest a solution to the energy shortage. The public thinks that domestic oil supplies are available and that the United States should not continue to rely on foreign energy sources. Finally, the last question in the table shows the results of a question concerning safety. Half of the public thinks that we can have offshore oil drilling and provide adequate environmental safeguards at the same time.

We may summarize these results as follows: large numbers of people (a) think energy shortages are serious threats, (b) do not want to solve them by importing foreign oil, (c) think the problem can be solved by drilling more oil in U.S. coastal waters, and (d) think coastal drilling is reasonably safe. Yet a substantial majority opposes further oil drilling. The safety question is critical here. Fifty percent said that further oil development was safe, yet only 34 percent supported it. That is, a substantial number of people (16 percent) believe that further oil drilling is safe, but undesirable—at least along the California coast.

Other solutions to the energy problem are possible, of course, but we have no evidence in our survey data to show which ones, if any, Californians regard as desirable. Additional drilling for oil in "government parklands and forest reserves" is rejected by majorities. Nuclear power is equally unpopular. Solar-power research was supported in the late 1970s and early 1980s, but about half of the Field Poll respondents agreed that "getting solar energy to a point at which it can generate substantial portions of our electrical power needs will take at least twenty years."[26] Moreover, as discussed in chapter 2, solar power certainly contributes to the nation's energy needs, but the contribution is fairly small and is not likely to grow rapidly in the near future. Conservation is also certainly an option, but as we saw in chapter 2, energy use per capita has been rising—despite people's claims that they are environmentalists and that they want to conserve energy. In other words, the other solutions are either unpopular or impractical.

There is only one obvious solution to these inconsistencies: Californians may simply want to have the benefits of plentiful, inexpensive oil and other forms of energy without bearing the costs of producing it in California. Indeed, we saw anecdotal evidence of this in chapter 1. We also have some evidence of that from a student classroom survey conducted at the University of California at Santa Barbara. This survey showed that the students were more favorably inclined toward oil drilling in remote areas of the California coast than they were toward oil drilling in Santa Barbara County (a view contrary to the views of most environmentalists, as well as policies set by the California Coastal Commission).[27] Unfortunately, we have no general population survey data about Californians' attitudes toward increased oil production in such places as Alaska, Louisiana, or Texas, or in even more distant areas. So we cannot say with certainty that Californians would be more supportive of oil development elsewhere. Nevertheless, though the data in table 4.4 do not prove that some people's thinking is inconsistent, they certainly suggest it.

Attitude Consistency

Although our tour of the opinions described in table 4.4 offers some tantalizing insights suggesting poorly thought-out opinions, the analysis is not conclusive. Because the questions did not eliminate all logical possibilities, they could not show logical inconsistency. We move on now to another approach to consistency, one that does not rely on the high standard of logic, but that is nevertheless quite useful for understanding how people think about politics—a study of attitude consistency.

Attitude consistency is the degree to which one's opinions on political issues are all roughly at the same point on the ideological spectrum. In terms of environmental issues, if one holds all strong environmentalist positions, or all centrist positions, or all strongly pro-development positions, one would be considered "consistent." In contrast, if one holds a mixture of pro-environment, centrist, and pro-development opinions all at the same time, then one would be considered "inconsistent." Although having a consistent set of opinions does not necessarily mean that one is a sophisticated thinker about politics, attitude consistency and sophistication are closely associated. Most public opinion scholars regard low attitude consistency as a sign that people have not thought much about the issues in question and that their thinking is muddled and confused.[28]

It may help to describe attitude consistency in more concrete terms. Consider the example of someone who agrees with the two following statements from an October 1981 Field Poll:[29]

- "It would be better to cut back on living standards in order to conserve energy rather than to go on using up natural resources at the present rate."
- "To encourage more exploration and drilling for natural gas in the United States, the price of natural gas should be completely decontrolled."

Although, strictly speaking, the two statements are not logically contradictory, common sense tell us that one should not agree with both. If one favors cutting back on living standards to conserve natural resources, one should not also favor changing laws that effectively encourage using up natural resources even faster. Those two views are at the opposite ends of the political spectrum on environmental issues—the first stand favors conservation, the second favors development.

Not only are these two views in conflict, but when the questions were asked in 1981, energy policy was the focus of a great deal of attention. The last year and a half of Jimmy Carter's presidency was dominated by the Iranian hostage standoff and the second energy crisis. Gasoline prices shot up. Gas lines reappeared at gas stations. Energy policy proposals—including decontrol of oil and natural gas prices—were regularly covered by the news media during the oil crisis and the presidential campaign. Republican presidential nominee Ronald Reagan believed in the free market and attacked Jimmy Carter's failure to tame the energy beast during the campaign. Reagan conspicuously declared that one of his first official acts as president would be to decontrol oil prices. In early 1981 he carried out the promise, and the debate turned to decontrol of natural gas prices. So for survey respondents in 1981, more than common sense was telling them that the opinions expressed in the two statements were in conflict.

Politicians were telling them as well. Environmentalists were calling for conservation; their opponents were calling for developing new energy supplies.

In other words, people who paid attention to politics and thought about energy issues should have understood that cutting back on living standards to reduce energy use and decontrolling natural gas prices to encourage exploration and use were opposing alternatives. If people were inconsistent, it was because they did not think much about the issues and did not learn enough to follow the basic policy debates during the 1980 presidential campaign. That is why attitude consistency is so closely associated with political knowledge. Knowledgeable people who pay attention to politics are generally consistent; uninformed people who ignore politics are generally inconsistent.

To explore the attitude consistency of the California public on energy issues, we can examine responses to a number of questions about energy policy asked by the Field Poll in two surveys of Californians in 1977 and 1998.[30] These two years offer us a nice range of history. Not only do they span over twenty years, but the first survey was conducted when energy issues were high on the nation's agenda, while the last was conducted during a time when energy issues generally received little attention.

The subjects of the questions differed slightly from one year to another, to address current policy issues (see the data appendix for exact question wordings). For example, as discussed in chapter 2, whether to allow the construction of liquefied natural gas (LNG) tanker terminals was a major policy question in 1977, but not in 1998. Yet the fact that the questions changed is not important to our central findings, which deal with how people understand various energy and environmental policies and how coherently they think about related policy issues.

We should also note that the sorts of questions here differ sharply from the questions typically used in analyses of belief systems and attitude consistency. Conventional investigations of belief systems use questions that tap a wide range of important issues of the day—what Philip Converse once described as a "purposive sampling" of salient issues. Most studies, in fact, exclusively rely on the questions used by one of the major academic public opinion surveys—the CPS/SRC American National Election Studies. The researchers designing the ANES surveys write questions to assess people's responses to the major election issues of the day; consequently, the questions in any single survey range widely, covering economic, social, and foreign policy. These surveys rarely ask series of closely related questions, and when they do (for example, on racial integration), the answers are generally not used in analyses of belief systems. Moreover, the questions broadly generalize about policies, rather than ask about specifics. For example, in the 1980s the NES survey asked whether the United States should cooperate generally with the Soviet Union, rather than about a specific instance of cooperation, such as building an oil pipeline from the Soviet Union to the West to increase trade. Even when researchers use other surveys, they almost always ask a wide range of general questions about politics, in the style of the ANES surveys, rather than a series of related questions about a single policy area.[31]

This analysis, in contrast to others, uses a series of closely related questions about specific energy policies. A few of the questions in the later surveys are about general environmental principles, but most are about concrete policies. Public debates about energy policy addressed all of these policy proposals. Indeed, the Field Poll asked these questions because the answers were newsworthy. Legislators in Congress and the California State Legislature eventually enacted some of these ideas into law. Policy makers and members of related attentive publics (if any) would certainly regard the relationships among these policies as obvious.

The choice of questions for analysis leads to two alternative hypotheses. First, the relationships among the attitudes might be very strong because the issues are so closely related. Second, the relationships among the attitudes might be very weak because the issues deal with practical policies and are therefore complicated and poorly understood.

Before wading through the data, let me preview the main finding. Examining attitudes toward a range of energy policies shows a surprisingly low level of consistency. In other words, people do not consistently take pro–energy development stands or pro–energy conservation stands. The level of consistency is so low that we must conclude that a substantial number of people must have what Philip Converse called "non-attitudes."[32] That is, a substantial number of people must be thinking about the problem posed in the survey question for the first time when the survey interviewer asks the question. Previous research suggests that when people are asked questions regarding issues about which they know little, they respond in an almost random fashion, so that their opinions change from one interview to the next, and so that their opinions on one issue are only weakly related to their opinions on others.[33] That is what we see here.

These findings do not mean that people have no basis at all for their opinions. Even poorly informed people apparently draw on some factual and value-laden considerations when they answer survey questions.[34] Moreover, the ambiguity in the questions causes some of the apparent randomness in answers.[35] Nevertheless, the data that we will see in the following tables indicate that many people are likely to be quite poorly informed about, and uninterested in, many of the energy questions used in our analysis.

Attitude consistency is normally assessed using some measure of association. These measures summarize the strength of a relationship in a single number. The measures used here, Pearson product-moment correlation coefficients, range from − 1 to + 1. Values of plus or minus 1.0 indicate perfect relationships, while a value of zero represents no relationship whatsoever. That is, if the correlation is 1.0, all respondents give exactly the same answers to a pair of attitude questions. For example, if one gave a strongly environmental answer to one question, he or she would also give a strongly environmental answer to the other question. Consequently, a researcher could perfectly predict a respondent's answer on one environmental question on the basis of his or her answer to the second question. The closer the value of the correlation is to zero, the less able the researcher is to predict one answer from another. When the value is zero, knowing that a respondent gave a strongly environmental answer to one question gives the researcher no basis for guessing what the respondent's answer would be to a second question.[36]

Let us turn to the data. Table 4.5 presents a set of Pearson correlation coefficients among fourteen questions asked in 1977 about energy policies designed to help alleviate shortages. All the questions ask whether respondents agree with particular policies; there are no general questions about support for environmental principles or symbols. In the eyes of policy makers, some of the policies are pro-development, in that they seek to increase the energy supply (for example, drilling or importing more oil or liquid natural gas, building nuclear power plants, using more coal); other policies are pro-conservation, in that they seek to reduce energy consumption (for example, tax credits for insulating homes or installing solar energy, improving mass transportation, reducing living standards). Of course, the policies are not mutually exclusive, but there is little doubt that policy makers see pro-development and pro-conservation sides on each issue, and that they see them as opposites.

The first observation we can make is that these correlations are almost all quite small. Forty of the ninety-one correlations are less than .10 in magnitude—substantively trivial. In fact, thirty-one of the correlations are not even statistically significant at the .05 level. In other words, they are so small that they may differ from zero only by random sampling error. On average, these correlations are far weaker than correlations that led Converse to his famous conclusions about nonattitudes.[37]

Converse, of course, looked at major issues of the day, many of which were not very closely or substantively related to one another (for example, whether the federal government should provide aid to education, and whether it should keep soldiers abroad to resist communism). This led to our first hypothesis, that items closely related on substantive grounds would be more highly correlated. Here we see that the data do not support this hypothesis. The reason seems to be that these questions are all about what Carmines and Stimson called "hard" issues.[38] That is, these issues require some minimal knowledge of economics and energy policy, and they oblige the respondents to think critically about them.

Table 4.5 Correlations among Energy Attitudes in 1977

	A	B	C	D	E	F	G	H	I	J	K	L	M	N
A	1.00													
B	0.31	1.00												
C	0.30	0.73	1.00											
D	0.23	0.13	0.16	1.00										
E	0.18	0.20	0.26	0.16	1.00									
F	0.16	0.08	0.11	0.14	0.16	1.00								
G	0.01	0.03	0.13	0.07	0.14	0.06	1.00							
H	0.06	0.02	0.07	−0.02	0.09	0.13	0.11	1.00						
I	0.03	0.00	−0.01	0.07	−0.07	−0.05	−0.03	−0.07	1.00					
J	−0.14	−0.03	−0.05	−0.06	−0.07	−0.11	−0.02	0.10	0.12	1.00				
K	−0.09	−0.05	−0.08	−0.08	−0.07	−0.12	−0.06	0.11	0.19	0.23	1.00			
L	−0.10	−0.06	−0.05	−0.06	−0.05	−0.13	−0.02	0.13	0.16	0.19	0.52	1.00		
M	−0.07	−0.05	−0.05	−0.04	−0.06	−0.14	−0.03	0.09	0.16	0.15	0.39	0.52	1.0	
N	−0.08	−0.07	−0.07	−0.05	−0.09	−0.04	−0.05	0.08	0.18	0.18	0.34	0.34	0.34	1.0

Variables

A: Tax low-gas-mileage cars
B: Tax credit for insulating homes
C: Tax credit for solar energy in homes
D: Raise gas prices to reduce driving
E: Improve mass transportation
F: Cut back on living standards to conserve energy
G: Give solar energy more priority
H: Control prices, but give tax breaks for oil exploration
I: Decontrol oil and gas price to encourage exploration
J: Cut air pollution standards & substitute coal for oil and gas
K: Drill more oil in tidelands along coast
L: Build more terminals for unloading oil
M: Build more terminals for unloading liquid natural gas (LNG)
N: Build more nuclear power plants

Source: Data are from the Field Institute's California Poll 7703, 17 June–2 July 1977.
Note: Correlations are Pearson product-moment correlation coefficients.
Minimum sample size for correlations = 733.

The questions do not have simple subjects or symbols that allow respondents to take positions on the basis of simple emotional responses.

More generally, we can speculate that this pattern of poorly related responses will be characteristic of any set of detailed questions about policy options. As Fischhoff, Slovic, and Lichtenstein put it, "When faced with complex, unfamiliar issues, people may have poorly formulated, even incoherent values."[39] We are seeing a large dose of incoherence here. People may know what they want at the level of general environmental principles, but they are less likely to know what those principles imply in terms of practical policies.

Looking in more detail at the correlations, we see that two clusters of correlations stand out. The questions about taxing low-gas-mileage cars and tax credits for insulating or using solar energy (items A, B, and C), and to a lesser extent the questions about raising gas taxes and improving mass transportation (items D and E), all form one cluster. That is, these items correlate more highly with one another than with questions that are not part of this cluster. A second cluster is formed by the questions about drilling more offshore oil, building seaport terminals for importing oil and liquid natural gas, and building nuclear power plants (items K, L, M, and N). The questions about decontrolling the price of oil and gas, cutting air pollution standards, and using coal also seem to fit into this cluster, although more weakly. In broad terms, we can describe the first cluster as favoring conservation and the second as favoring development.

The key observation about these clusters is that they are barely related to one another. The conservation policies are negatively correlated with the development policies, which is the correct direction for the correlations. That is, the more one favors conservation, the less one should favor development, and vice versa. But the correlations are substantively trivial, ranging from .05 to .10. In other words, the evidence suggests that the public does not view conservation and development policies as opposites. Policy makers may regard these policies as competing alternatives, but most Californians in 1977 did not seem to think of them in that way. They apparently saw neither that the conservation and development policies are alternative means of reducing the energy shortage nor that there are relationships among the fourteen individual policies.[40]

The 1998 correlations, presented in table 4.6, differ slightly from the previous data. The correlations are still generally weak in 1998, but not as weak as the 1977 correlations. Only five of the twenty-eight correlations are less than .10 in magnitude. In addition, although the pro-energy development opinions (nuclear power and more drilling on the coast or in parklands) stand out from the energy conservation opinions (increasing energy taxes, cutting living standards, and slowing growth), the differences are not as sharp as they were in 1977. The clusters are not quite as distinct. We will return to a discussion of the differences

Table 4.6   Correlations among Energy Attitudes in 1998

	A	B	C	D	E	F	G	H
A	1.00							
B	0.32	1.00						
C	0.27	0.13	1.00					
D	0.17	0.12	0.19	1.00				
E	0.27	0.14	0.29	0.35	1.00			
F	−0.15	−0.03	−0.26	−0.18	−0.19	1.00		
G	−0.19	−0.10	−0.20	−0.02	−0.02	.36	1.00	
H	−0.03	−0.03	−0.15	−0.11	−0.21	.22	.12	1.00

Variables
A: Increase energy taxes to get business to use energy more efficiently
B: Extra tax on low-gas-mileage cars
C: Cut back on living standard to conserve energy
D: Slow population and housing growth
E: Slow industry growth
F: Drill more oil in tidelands along coast
G: Drill more oil in parklands and forests
H: Build more nuclear power plants

Source: California Offshore Oil Drilling and Energy Policy Survey, March 1998, conducted by the Field Institute.
Note: Correlations are Pearson product-moment correlation coefficients.
Minimum sample size for correlations = 734.

between 1977 and 1998 after looking at more evidence on the patterns within each year.

In both surveys, the correlations between any two energy development questions, or between any two energy conservation questions, are higher than the correlations between a development question and a conservation question. The differences are not large, but they indicate that the public does not see development and conservation as opposites—contrary to the views of many interest groups, policy makers, and other political elites.

A factor analysis of the policy questions provides more support for this conclusion. If the analysis were to show a single factor causing the observed responses to the questions, we would have to conclude that the unobserved factor was attitude toward environmental issues. It would follow that pro-conservation and pro-development attitudes were opposites. In contrast, if the analysis were to reveal two factors—one causing responses to the energy development questions and a second causing responses to the energy conservation questions—we would have to conclude that a large portion of the public does not regard energy development and conservation as opposites. As we shall see, our data support the second conclusion.

Table 4.7 presents a factor analysis of the 1977 data. The key question here

Table 4.7   Maximum-Likelihood Factor Analysis of 1977 Energy Attitudes

Variable	Factor 1	Factor 2
Build more oil/gas terminals	−.18	.74
Build more nuclear power plants	−.17	.52
Reduce clean air standards/use coal	−.17	.28
Build more LNG terminals	−.15	.60
Drill more coastal oil	−.13	.63
Decontrol oil & gas prices	.03	.27
Tax breaks for oil exploration	.05	.11
Give solar energy high priority	.13	−.04
Cut living standards to conserve	.16	−.18
Raise gas taxes	.24	−.07
Improve mass transportation	.30	−.10
Tax low-mileage cars	.35	−.08
Tax credit for insulation	.78	.10
Tax credit for solar energy	.90	.11
Eigenvalue	6.56	3.26
Percent variance explained	.57	.23
Sample N = 462		

Note: Varimax rotation
Source: Data are from California Poll 7703, 17 June–2 July 1977.

is, how many factors are there? The usual tests for number of factors—size of eigenvalues and a scree test—show that a two-factor solution fits the data best. This finding supports the hypothesis that pro-conservation and pro-development views are not opposites. To explore this possibility farther, we need to examine the details of the factor loadings.

The variables in table 4.7 are sorted in order of their loading on factor 1 to help interpret the data. The two tax-credit items dominate the first factor, with loadings of .78 and .90. If we were to follow the traditional rule of thumb and ignore loadings less than .40, we would have to stop with those two items. Taking a broader perspective, we can say that improving mass transportation and taxing low-gas-mileage cars also characterize this factor. By whatever standard, the factor seems to reflect conservation policies. We can interpret the second factor more easily. Four items have loadings greater than .40—building more oil and gas terminals, building more nuclear power plants, building more liquefied natural gas terminals, and drilling more coastal oil. This factor reflects policies favoring development. As in the correlational analysis above, we find that conservation and development policies were not seen as opposites in 1977. Rather than forming a single factor, they load on different factors—providing evidence that the public did not view conservation and development as opposing energy policies.

The 1977 data are not a fluke. A factor analysis of a 1981 survey (data not shown) yields a pattern similar to that found with the 1977 data. A two-factor model clearly fits the data better than any other. Again, conservation items dominated one factor, while development items dominated the other factor. So the evidence suggests that during the energy-crisis years, Californians did not think of energy policy in simple terms of conservation versus development.

The 1998 data, shown in table 4.8, yield a somewhat different picture. Here a single-factor solution fits the data best. As our examination of the correlations in table 4.6 suggested, the issue clusters are not sufficiently distinct to fall into different factors. Although the public seemed to have viewed conservation and development as barely related in the energy-crisis years, by 1998 they seem to have recognized that conservation and development are opposites. Or did they?

There are three possible explanations for the apparent change in how people think about energy-policy questions. The first explanation is that opinions could have changed. After years of reading and hearing about environmental policies, the public might have come to think of the choices in the same way their leaders do. Environmentalists want to conserve energy rather than build more nuclear power plants or drill more oil along California's coast, while those on the pro-development side want the opposite. Perhaps the public now thinks in the same way. The second explanation is that the apparent change in the public's response could merely reflect the different sets of questions asked each year. Our surveys, after all, do not have identical questions. Perhaps the 1998 survey included questions that were more strongly related to one another and that related along a single, conservation-versus-development dimension. The third explanation is that people's knowledge and understanding of energy issues could have faded

Table 4.8   Maximum-Likelihood Factor Analysis of 1998 Energy Attitudes

Variable	Factor 1
Drill more coastal oil	−.45
Drill more oil in parks/forests	−.31
Build more nuclear power plants	−.28
Tax low-mileage cars	.30
Slow population/housing growth	.43
Increase energy taxes on businesses	.50
Slow industrial growth	.53
Cut standard of living—conserve energy	.54
Eigenvalue	1.92
Percent variance explained	.24
Sample N = 601	

Source: California Offshore Oil Drilling and Energy Policy Survey, March 1998, conducted by the Field Institute.

over time, causing even the environment and development clusters to break down. That is, perhaps as people paid less attention to energy issues, they became less likely to recognize that various pro-conservation policies were related and, similarly, that various pro-development policies were related. Instead of growing knowledge, we might be seeing declining knowledge causing opinions to become haphazard and chaotic.

The way to sort out what happened is to look at an identical set of questions over time. Table 4.9 does just that. It shows all the correlations between pairs of survey questions that were asked both in 1998 and in either the 1977 or 1981 survey. The upper-right half of table 4.9 shows the correlations from the 1998 survey. The lower-left half shows the same correlations from the two earlier surveys. Correlations that are larger in magnitude than their counterparts in the other half of the table (that is, in different years) are in bold type. These comparisons help us see what happened.

Six of the twelve correlations from 1977 and 1981 differ in strength from their 1998 counterparts. In every case of change, the earlier correlations are stronger. For example, the correlation between opinions on nuclear power and offshore oil drilling was 0.41, but by 1998 it had shrunk to only 0.22. Similarly, the correlation between opinions on nuclear power and oil drilling in state parks and forests was 0.28 in 1981, but by 1998 it was only 0.12. Collectively, these changes suggest—as the third explanation claims—that the public's understanding of energy policy has actually been diminishing. They are less able now than twenty years ago to recognize what policies consistently fit together. They are less likely to know, for instance, that good environmentalists should oppose both

Table 4.9 Selected Correlations among Energy Attitudes, 1977–1998

	A	B	C	D	E	F
A. More nuclear power	1.00	.22	.12	.03	−.21	−.11
B. Drill coastal oil	**.41**	1.00	.36	−.03	−.19	−.18
C. Drill park/forest oil	**.28**	.42	1.00	−.10	−.02	−.02
D. Tax low-mileage cars	**−.08**	−.09	*	1.00	.14	.12
E. Slow industrial growth	−.29	−.17	**−.11**	*	1.00	.35
F. Slow population growth	−.13	−.13	**−.11**	*	**.45**	1.00

Entries above the main diagonal (in *italics*) are 1998 data.
Entries below the main diagonal are 1977 or 1981 data.
Entries in **bold** font are at least .08 larger in magnitude than the corresponding entries for the other years.

Source: Data are from the Field Institute's California Poll 7703, California Poll 8104, and the California Offshore Oil Drilling and Energy Policy Survey, conducted by the Field Institute.
Note: The tax low mileage cars question was asked in 1977, but not 1981. All correlations using that item are from 1977. All other correlations below the main diagonal are from 1981.

nuclear power and further offshore oil development, or that good pro-development people should support them.

In connection with this evidence of the public's declining understanding of energy policy, it is probably also worth mentioning the decline in the ability of the factor-analysis models to explain people's opinions. As table 4.7 shows, the first factor in the 1977 model explains 47 percent of the variance, and the second explains an additional 23 percent—for a total of 70 percent. In 1981, the first factor explains 28 percent of the variance, and the second factor adds 11 percent—for a total of 39 percent. In our 1998 data set, however, only one factor appears, and it explains only 24 percent of the variance in the energy questions. The trend across the three surveys is that the underlying factors explain less and less of the observed variables. To put it another way, over the years people's opinions became less strongly related to one another—evidence that the public's understanding of energy policy was fading over time. This evidence is not conclusive because the questions in the factor analyses changed from one survey to the next, but it adds to the case that what we are seeing is a fading of the public's understanding of energy policy.

A FEW IMPLICATIONS

At this point, it would be useful to pause and draw out the implications of these findings about knowledge and attitude consistency. Previous studies about environmental attitudes have mostly examined one attitude at a time, leaving the impression that the public has at least a moderate understanding of environmental issues. The results here paint a different picture. In general, the public does not seem to know much about energy policy. Not only do most people lack factual knowledge about energy issues, most people hold opinions that do not fit any consistent worldview of energy and environmental issues. Our examination of attitude consistency did not reveal any evidence that many people thought about energy issues from a consistently pro-environmental perspective, nor that many people consistently favored development. Instead, we saw a public in which most people took an almost random mix of pro-environment and pro-development stands. In the 1977 and 1981 surveys, respondents did not seem to regard energy conservation and energy development—or what some writers and policy makers think of as pro- and anti-environment stands—as being opposites. Most people call themselves "environmentalists," but they do not consistently hold opinions that go with that label.

Of course, energy development and energy conservation are certainly compatible, and perhaps we should not criticize people for favoring policies that some might consider "balanced approaches."[41] Yet the evidence about lack of knowledge implies that the public's views are not so much balanced as confused and uninformed. Most people apparently value the environment and want to protect it but do not understand what government policies follow from that desire.

Another implication is that most people in the world of policy making— legislators, lobbyists, bureaucrats, and political activists—see environmental disputes in ways quite different from how the general public views them. Policy makers see most pro-environment and pro–energy development policies as ideologically conflicting approaches to solving our energy problems. To put it in terms of a current example, some policy makers think that global warming is a serious problem and that we ought to reduce fossil-fuel use and conserve energy in order to minimize it. Others think that scientists have not shown that global warming is a problem and that we should be concerned with developing new energy sources to keep our economy growing. Rarely does one find a policy maker who both fears global warming and wants more oil development, or one who dismisses global warming as a myth but wants to reduce energy consumption by slowing growth. Yet in the public, one finds many people who favor such seemingly conflicting positions.

One way to look at the finding that the public and our political leaders see energy issues differently is that leaders have failed to explain the issues to the public. The Sierra Club and other environmental interest groups, for example, have done an excellent job of persuading the public that they ought to oppose nuclear power and oil development along the California coast. What they have not done is persuade the public to adopt strong policies to conserve energy. As the tables above show, someone who opposed nuclear power or oil drilling along the California coast had only a little more than a fifty-fifty chance of favoring or opposing any of the energy-conservation policies asked about in the Field surveys. Apparently most people do not make the connection that if we are not going to produce energy, we need to conserve the resources we have. This is a failure of education.

EDUCATING THE PUBLIC?

Our findings about the public's misinformation and generally low level of knowledge about energy issues lead to questions regarding what can be done to increase the public's knowledge. Can anything be done to raise the level of the public's understanding of energy issues? More practically, could a public-information campaign run by government agencies, oil companies, or environmental groups increase the public's understanding of the issues? To put a political spin on the question, could a public-information campaign manipulate public opinion by persuading the public to believe its sponsors' version of the "facts"?

Would it matter who was behind the attempt to educate the public? Although data are limited in this area, two sets of questions asked in the 1998 survey allow us to investigate how the public might respond to various sorts of campaigns.

These questions build on research in the psychology of persuasion. Social psychologists have long known that whether people accept persuasive messages depends partly on the credibility of the source of the message. For example, most people would be far more likely to believe a university scientist's findings about tobacco causing cancer than a tobacco company scientist's contrary findings because tobacco companies are not widely trusted—that is, they are not credible sources.[42] A second, less-well-established finding is that the content of the message may influence whether people accept it. Some studies have found that a message with "discrepant" information—information that contradicts one's prior beliefs—is less acceptable than a message that supports one's beliefs. According to this line of reasoning, people who believe that smoking causes cancer would be more likely to believe a new study showing that cigar smoking causes throat cancer than a study showing that it does not.[43] Other studies have searched for this message effect and failed to find it, so it remains a controversial hypothesis. Nevertheless, we will investigate both hypotheses in the context of oil-development policies. In the first case—communicator credibility—we are merely looking to see how the effect plays out in the case of a public-relations campaign about oil development; in the second case—message content—we are seeking to discover both whether evidence for the effect exists and whether it favors pro- or anti-oil political forces.

The first set of questions varies the source of scientific claims and seeks to show how the public might respond to news reports from government, oil industry, or environmental groups. The critical variable here is, which group's scientists does the public believe? In these questions, respondents were asked: "How much confidence do you have in statements made by [government/oil industry/environmental group] scientists about potential health risks associated with living near an oil drilling site? Do you have a great deal of confidence, a moderate amount of confidence, only some confidence, or almost no confidence at all?"

The results of these questions, shown in figure 4.2, indicate that environmental-group scientists are far more believable than government scientists, who in turn are more believable than oil-industry scientists.[44] These differences are both large and statistically significant.

The second set of questions varies the content of the scientific claims and seeks to discover if pro- or anti-environmental content influences whether a message is accepted. In other words, it allows us to see how the public might respond to news reports that were either pro- or anti-oil. Two questions were asked of randomly selected halves of the sample. One version of the questions

Figure 4.2   Having a "Great Deal" or "Moderate Amount" of Confidence in Scientists from Various Groups

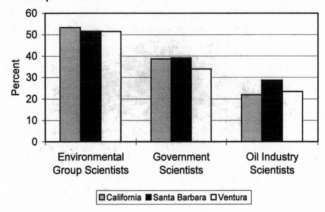

N = 810

said that offshore oil drilling is much safer than previously believed, and the other version that it was much riskier: "A group of university scientists recently declared that because of new technology, offshore oil drilling is much [safer/riskier] than previously thought. How much confidence do you have in this claim—a great deal of confidence, a moderate amount of confidence, only some confidence, or almost no confidence at all?"

The results, shown in figure 4.3, indicate that the California public is much more inclined to believe reports critical of the oil industry than reports favorable to it.[45] This difference is both large and statistically significant.

Curiously, Santa Barbara and Ventura samples differ from samples drawn from the state as a whole. Santa Barbara residents are less influenced by the content of the reports than Californians in general, and Ventura residents do not seem any more likely to accept pro-oil findings than anti-oil findings.

Taken together, the results shown in figures 4.2 and 4.3 indicate that the public is more likely not only to have environmentalist preferences but to believe research that supports the environmentalist position over the oil companies' arguments, and that they are more likely to believe environmentalist-group scientists than either government or oil-industry scientists. We began our look at educating the public by asking whether anything could be done to raise the public's understanding of energy issues. The answer is probably "Yes," but the effectiveness of any public education campaign will depend on which side is trying to educate the public and what the messages are. In any public debate, the public begins listening with the presumption that the environmentalists are more likely to be telling the truth than their opponents and that environmental claims are

Figure 4.3   Having a "Great Deal" or "Moderate Amount" of Confidence That Oil Drilling Is Safe or Risky

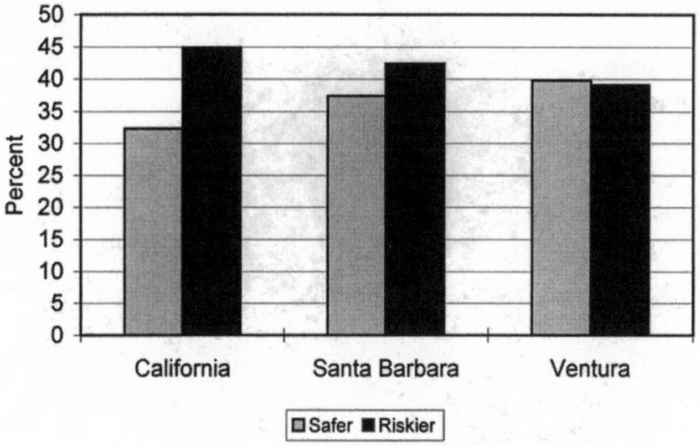

"Safer" question, N = 407; "Riskier" question, N = 403

more likely to be true than rival assertions. In other words, when both pro-environment and pro-development groups launch public-information campaigns, the environmentalists are far more likely to win the hearts and minds of the public.

CONCLUSION

In broad summary, the data in this chapter show that the public is not well informed about energy issues. The fact most critical to our energy policy—that the United States must import oil to meet its energy needs—was not known by about half the public during the energy crises of the 1970s. Many other facts were similarly unknown or misunderstood. Had other, more technical questions been asked, the public would presumably have been even more at a loss to answer. When we consider practical energy-policy questions facing current decision makers—such as whether the United States should move away from petroleum toward nuclear power in order to reduce greenhouse gases and global warming—we see that the public will clearly have a hard time even grasping what the debate is about, let alone understanding the arguments for and against various policy options. In their insightful study of global warming, based on a series of semistructured interviews and small samples of select groups, Willett Kempton and his colleagues found that many of their respondents confused global warming with either pollution or ozone depletion.[46] Although their study does not offer any large-scale, representative survey data to support their findings, the confusion they discovered is exactly what we should expect.

What should we conclude from these observations about the relationship between the public's knowledge of energy issues and their opinions on them? Most members of the public are not reasoning out their positions on the basis of careful assessments of the facts. They cannot, because they do not know many facts. Instead, people rely on general predispositions and opinion leadership. When people think about offshore oil development, for example, they do not weigh the economic benefits of increased oil production and more jobs against the air pollution and potential for oil spills or other harmful effects. Rather, they form opinions based most likely on a few scraps of information or on what a friend, a trusted politician, or a movie star says. In short, most people's opinions on energy issues are likely to be based on mental shortcuts and superficial judgments, rather than on substantial knowledge and careful reasoning. Public opinion on energy issues is certainly important and influential, but it is not necessarily well thought out.

NOTES

1. Because California does not produce any coal and because it is not a political issue in the state, I will not be looking at coal.

2. Peverill Squire et al., *The Dynamics of Democracy* (Madison, Wis.: Brown-Benchmark, 1997), 171; Michael X. Delli Carpini and Scott Keeter, *What Americans Know about Politics and Why It Matters* (New Haven, Conn.: Yale University Press, 1996), 117.

3. Delli Carpini and Keeter, *What Americans Know about Politics,* chap. 4.

4. The use of content analysis of news media reports has recently come under scrutiny because claims based on them were often not justified. Happily, the principal critic of the use of content analysis regards the sort of analysis I offer here as "straightforward" and without problems. See John T. Woolley, "Using Media Reports as Indicators of Policy Processes," *American Journal of Political Science* 44 (2000): 156–73.

5. These data are a simple count of the number of entries in the *Los Angeles Times* indexes under the heading "Petroleum and Petroleum Industry." The data presumably contain errors because different indexers may have made different indexing decisions in various years. The data begin in 1972 because that is the year of the first index.

6. Stephen Erfle and Henry McMillan, "Determinants of Network News Coverage of the Oil Industry during the Late 1970s," *Journalism Quarterly* 66 (1989): 121–28; Stephen D. Reese, John A. Daly, and Andrew P. Hardy, "Economic News on Network Television," *Journalism Quarterly* 64 (1987): 137–44; David Wolverton and Donald Vance, "Newspaper Coverage of Proposals for Rate Increases by Electric Utility," *Journalism Quarterly* 64 (1986): 581–84; Robert H. Bohle, "Negativism as News Selection Predictor," *Journalism Quarterly* 63 (1986): 789–96; Gerald C. Stone, Barbara Hartung, and Dwight Jensen, "Local TV News and the Good-Bad Dyad," *Journalism Quarterly* 64 (1987): 37–44; Gerald C. Stone and Elinor Grusin, "Network TV as a Bad News Bearer," *Journalism Quarterly* 61 (1984): 517–23.

7. The Pew Research Center for the People and the Press, *TV News Viewership Declines,* 13 May 1996, 76.

8. The Field Institute is a nonpartisan, not-for-profit public opinion research organization established by the Field Research Corporation (550 Kearny Street, Suite 900, San Francisco, California 94108). The sample was a representative sample of 810 adult residents of the state, with oversamples of 209 Santa Barbara County residents and 204 Ventura County residents. The interviews were conducted 5–18 March. The survey was designed by the principal investigator, Eric Smith, and funded by the University of California Toxic Substances Research and Teaching Program. These data will eventually be released to the public and archived at the University of California's UCDATA, on the UC Berkeley campus. Neither of these organizations is responsible for the analysis or interpretation of the data appearing here.

9. Colin J. Campbell and Jean H. Laherrère, "The End of Cheap Oil," *Scientific American,* March 1998, 78–83.

10. For a list of the world's biggest oil spills, see Joanna Burger, *Oil Spills* (New Brunswick, N.J.: Rutgers University Press, 1997), 30.

11. Pipelines are safer than tankers (as measured by the volume of the spills), but the difference may not be as great as some people would suspect. In recent years, the number of tanker and pipeline spills per billion barrels of oil transported in U.S. waters is roughly equal, but the median size of tanker spills is about twice that of pipeline spills. See Cheryl McMahon Anderson and Robert P. LaBelle, "Comparative Occurrence Rates for Offshore Oil Spills," *Spill Science and Technology Bulletin* 1, no. 2 (1994): 131–41.

12. Anderson and LaBelle, "Comparative Occurrence Rates for Offshore Oil Spills."

13. Gramling, Robert. *Oil on the Edge* (Albany: State University of New York Press, 1996), 68.

14. Burger, *Oil Spills,* chap. 12. Contact with crude oil poses no health threat to the general public. In principle, someone could be harmed by ingesting oil, but the chances of someone going down to the coast and eating globs of oil that float ashore when an oil spill occurs are sufficiently slim to be ignored. The real dangers associated with offshore oil platforms are occupational safety hazards to oil workers.

15. Gabriel Almond, *The American People and Foreign Policy* (New York: Harcourt, Brace, 1950); V. O. Key, Jr., *Public Opinion and American Democracy* (New York: Knopf, 1961), 265, 282–85; Philip E. Converse, "The Nature of Belief Systems in Mass Publics," in *Ideology and Discontent,* ed. David Apter (New York: Free Press, 1964). More recent scholars who have subscribed to the idea of attentive publics include Jon D. Miller, Robert W. Suchner, and Alan M. Voelker, *Citizenship in an Age of Science: Changing Attitudes among Young Adults* (New York: Pergamon, 1980), and Jon D. Miller, *The American People and Science Policy* (New York: Pergamon, 1983).

16. John Zaller, "Analysis of Information Items in the 1985 NES Pilot Study," Report to the Board of Overseers for the National Election Studies, 1986.

17. Shanto Iyengar, "Shortcuts to Political Knowledge: The Role of Selective Attention and Accessibility," in *Information and Democratic Processes,* ed. John A. Ferejohn and James H. Kuklinski (Chicago: University of Illinois Press, 1990).

18. Delli Carpini and Keeter, *What Americans Know about Politics,* app. 4, 330.

19. Jon A. Krosnick, "Government Policy and Citizen Passion: A Study of Issue Publics in Contemporary America," *Political Behavior* 12 (March 1990): 59–92.

20. Krosnick has found evidence of attentive publics, but another scholar, using a similar

approach, has failed to find them. See W. Russell Neuman, *The Paradox of Mass Politics: Knowledge and Opinion in the American Electorate* (Cambridge, Mass.: Harvard University Press, 1986), 67–73.

21. Nick Welsh, "All's Well That Ends Wells?" *The Independent,* 14 March 1996, 23–25; Andrew LePage and Melinda Burns, "Mobil Files Clearview Application," *Santa Barbara News-Press,* 10 February 1995, A1; Richard C. Paddock, "Drilling Advance Rekindles Santa Barbara Oil Wars," *Los Angeles Times,* 5 December 1994.

22. This conclusion was verified with a series of OLS and logistic-regression equations, none of which indicated that being a Santa Barbara or Ventura resident gave people more than a tiny advantage in answering any of the information items correctly. In the one equation showing an effect, the three samples were combined (total n = 1,223), and an additive index of the four oil information questions was regressed on a set of independent variables that included dummy variables for Santa Barbara and Ventura. Being a Santa Barbara resident was of borderline significance (p < .06), raising one's score on the five-point index by 0.11—a substantively trivial amount.

23. Delli Carpini and Keeter, *What Americans Know about Politics,* 305–306. Two of the questions were slightly reworded. See the data appendix for details. Our data clearly show that the slight variations had no effect on their use as a measure of general information.

24. The Santa Barbara and Ventura samples are combined in table 4.3; however, I performed the factor analysis and all other tests in the two county samples separately before combining them. No substantive results or conclusions differed when the two samples were combined.

25. I estimated a series of Lisrel models with both California and Santa Barbara/Ventura data sets, using both tetrachoric correlation coefficients and covariances. The two basic models were (a) that a general factor caused all items and a second oil factor influenced only the oil-related items, and (b) that a general factor explained the five Delli Carpini–Keeter items and a second factor explained the oil-related items. I also estimated the models dropping the obviously badly fitting oil-import and spill-frequency questions. All of the models had serious fit problems. In the best-fitting model, the general and oil factors correlated at .91, but the fit was poor, and one of the parameter estimates was theoretically impossible. In sum, none of these analyses produced a shred of evidence to suggest that a second, oil-related knowledge factors exists.

26. Field Polls 7902 and 8002.

27. This survey is described and analyzed in chapter 5.

28. Converse, "The Nature of Belief Systems in Mass Publics"; Eric R. A. N. Smith, *The Unchanging American Voter* (Berkeley: University of California Press, 1989).

29. Field Poll 8104.

30. The polls were 7703 (June 1977) and the California Offshore Oil Drilling and Energy Policy Study (March 1998).

31. For a review of the literature, see Smith, *Unchanging American Voter,* chap. 4.

32. Philip E. Converse, "Attitudes and Nonattitudes: The Continuation of a Dialogue," in *The Quantitative Analysis of Social Problems,* ed. Edward Tufte (Reading, Mass.: Addison-Wesley, 1970).

33. Converse, "Nature of Belief Systems in Mass Publics" and "Attitudes and Nonattitudes"; Gillian Dean and Thomas W. Moran, "Measuring Mass Political Attitudes: Change and Unreliability," *Political Methodology* 4 (1977): 383–413.

34. John R. Zaller, *The Nature and Origins of Mass Opinion* (Cambridge: Cambridge University Press, 1992).

35. Christopher H. Achen, "Mass Political Attitudes and the Survey Response," *American Political Science Review* 69 (1975): 1218–31.

36. In the original research on attitude consistency, gammas were used because they only assume ordinal-level measurement. I use Pearson correlations here for two reasons. First, the substantive results are identical whether one uses gammas or Pearson r's. Second, I must assume interval-level measurement in the factor analysis elsewhere in this chapter.

37. Converse, "Nature of Belief Systems in Mass Publics."

38. Edward G. Carmines and James A. Stimson, *Issue Evolution: Race and the Transformation of American Politics* (Princeton, N.J.: Princeton University Press, 1989).

39. Baruch Fischhoff, Paul Slovic, and Sarah Lichtenstein, "Poorly Thought-Out Values: Problems of Measurement," in *Energy and Material Resources: Attitudes, Values, and Public Policy,* ed. W. David Conn (Boulder, Colo.: Westview, 1983), 39.

40. Analysis of an October 1981 poll (Field Poll 8104) revealed essentially the same findings.

41. For a study that takes this view, see Eugene A. Rosa, Marvin E. Olsen, and Don A. Dillman, "Public Views toward National Energy Policy Strategies: Polarization or Compromise?" in *Public Reactions to Nuclear Power: Are There Critical Masses?* ed. William R. Freudenburg and Eugene A. Rosa (Boulder, Colo.: Westview, 1984), 69–93.

42. Alice H. Eagly and Shelly Chaiken, "Cognitive Theories of Persuasion," *Advances in Experimental Social Psychology* 17 (1984): 267–359; Susan T. Fiske and Shelley E. Taylor, *Social Cognition* (New York: Random House, 1984), chap. 12.

43. G. Greaves, "Conceptual System Functioning and Selective Recall of Information," *Journal of Personality and Social Psychology* 21 (1972): 327–32; Fiske and Taylor, *Social Cognition,* 362–63.

44. The entire sample was asked these three questions. The order of the questions was rotated.

45. In figure 4.2, the respondents were randomly asked one of two versions of the question. Because randomization does not normally yield identical half-samples, the samples sizes are only approximately equal.

46. Willett Kempton, James S. Boster, and Jennifer A. Hartley, *Environmental Values in American Culture* (Cambridge, Mass.: MIT Press, 1995), chap 4.

5

What Causes Public Opinion?

The final step in our examination of public opinion toward energy issues is exploring who takes what side in disputes over energy policy. What are the patterns of support and opposition on oil development and nuclear power? How do opinions differ from group to group? Are people guided by self-interest or by other motives? Does it matter how close one lives to an oil-drilling operation or nuclear power plant? What role does ideology play in influencing opinions on energy issues? How have these relationships changed over time? Taken together, the answers to these questions will help us understand what causes public opinion on energy issues.

The task for this chapter is to explain the public's preferences on the main options for increasing energy production from conventional sources—by drilling for oil and gas offshore, by drilling in parks and public lands, and by increasing nuclear power production. In order to get an in-depth look at the problem, I will use a series of public opinion polls conducted in California. These surveys were designed specifically to investigate opinions on energy production in the state; they offer far more detail than available national surveys. In cases where comparable national data were available, however, the results basically matched the California results. In other words, California is a reasonable microcosm of the United States for purposes of studying public opinion on energy issues.

Broadly speaking, two approaches have been used to explain the public's views on environmental issues—the demographic approach and the ideological approach. In this examination of energy opinions, we will combine both approaches because the two complement one another. The combined result yields a more comprehensive explanation than either approach pursued alone.

The first approach is to rely on demographic variables—education, age, income, race, gender, and related variables—along with party identification and self-identified ideology to explain opinions. One might think it odd to include ideological identification with the demographic variables, but self-identified ideology is measured with a single question, rather than with a battery of questions typical of the second, ideological approach. Moreover, self-identified ideology, party identification, and the demographic variables all come from what one might think of as the public opinion researcher's standard tool kit.

The weakness of this approach is that it offers no grand, unifying public opinion theory to guide readers through the analysis. While we can explain variable by variable why we should expect to find particular relationships, we cannot offer any overarching scheme to explain the connections. We cannot, for example, return to "first principles" (as rational-choice theorists describe the foundations of their theory) and explain why young, well-educated, liberal Democratic women should be expected to be among the strongest opponents of nuclear power.

The strength of this approach is that the variables are recognized both by public opinion researchers and political practitioners. The variables I will examine include many that are used by politicians and activists when they plot lobbying, fund-raising, or campaign strategies. Some of the variables—political party, age, gender, and ethnicity, for example—are data that are available to every campaign from voter registration tapes.[1] Others are available in such sources as campaign polls or can be inferred (for example, one can infer from census data the income levels of people living in different neighborhoods). In short, what the demographic approach lacks in grand theory, it makes up in practicality in real-world politics.

The ideological approach, in contrast, focuses on deeply held values, or what some scholars describe as "cultural biases" or "worldviews." Multiple questions are used to construct indexes of "egalitarianism," "individualism," "postmaterialism," and other values. As such, these measures go well beyond the simple question, "Are you a liberal, a moderate, or a conservative?" In doing so, they attempt to tap into people's basic values and show how they influence people's opinions on the issues of the day. Several sets of scholars have pursued alternative versions of this search for deeper values to explain environmentalism. I will investigate two of the most prominent ideological constructs—cultural theory and postmaterialism theory.[2]

The strengths and weaknesses of the ideological approach are a mirror image of the strengths and weaknesses of demographic approaches. On the positive side, there are clearly articulated principles and arguments to support these theories, and they do a better job of explaining people's opinions than do demographic variables. On the negative side, ideological measures are rarely included in public opinion surveys. Moreover, they are not likely to be picked up by politicians or political activists for use in their struggles. One can target mass mailings by age or gender; egalitarianism, however, is not included on voter registration tapes.

At the end of this chapter, I move beyond existing studies to propose and test a more comprehensive explanation of people's attitudes toward oil drilling and nuclear power—and, indeed, toward any potentially risky environmental technology. The theory proposed here combines two existing theories: Mary Douglas and Aaron Wildavsky's theory of cultural worldviews and John Zaller's theory explaining how people acquire their attitudes from the news media and mass communications campaigns. Working together, these theories give us a far better understanding of public reactions to energy and environmental issues than does any previous attempted explanation.

This chapter begins by describing the questions used to measure public opinion toward oil drilling and nuclear power, and setting forth the historical context of the surveys in which the questions were asked. I then turn to a brief review of research explaining how people learn about new issues and forget old ones. A good deal of attention has been given to how people learn about new issues as they arise. Scholars like John Zaller, Benjamin Page, and Robert Shapiro have developed strong theoretical insights explaining how the public reacts to new issues.[3] Relatively little attention has been given to explaining what happens to people's opinions as issues fade from the front pages, as energy issues have in the late 1990s; I will consider what we should expect to see in such cases. I then proceed to examine how opinions on energy development relate to a series of variables taken one at a time—following first the demographic approach and then the ideological approach. Finally, I will prersent a set of comprehensive multivariate models to explain opinions on energy issues.

QUESTIONS ABOUT ENERGY

To examine public opinion about expanding production from conventional energy sources, we use three questions asked in a series of public opinion surveys of California adults in 1981, 1990, and 1998—the same three questions examined in chapter 3.[4] These three years give us pictures of opinion at three very different times. By looking at these data, we can get a sense both of how opinion changed over time and how public opinion on these issues responds to different political situations.

The three questions asked respondents to agree or disagree with the following statements:

- "Oil companies should be allowed to drill more oil and gas wells in state tidelands along the California seacoast."

- "Current government restrictions prohibiting the drilling of oil and gas wells on government parks and forest reserves should be relaxed."
- "The building of more nuclear power plants should be allowed in California."

These questions are all phrased as agree-disagree questions, with agreement favoring more energy development. As explained in chapter 3, because some respondents display "acquiescence bias"—that is, the tendency to agree with any statement posed in an agree-disagree format—the questions probably slightly exaggerate the support for oil drilling and nuclear power. Still, they offer an opportunity to examine the views of different social groups and to learn who supports further energy development and who opposes it.

Although the questions remained the same over time, the historical context did not. When the Field Institute conducted the first survey in 1981, the second energy crisis was still upon us. The bloody Iran-Iraq War was taking a toll both in lives and in lost oil production. In the United States, although the gas lines were finally a frustration of the past, oil prices had reached their post–World War II high. Moreover, the Middle Eastern war and high gas prices were not the only energy policy issues to make front-page news that year. In the spring of 1981, President Reagan fulfilled one of his campaign promises by decontrolling oil prices. He followed that move with well-publicized—although unsuccessful—attempts to abolish the Energy Department and decontrol the price of natural gas. President Reagan's new interior secretary, James Watt, also focused attention on energy issues by attempting to open huge tracts of land for oil development—in the Gulf of Mexico, in Alaska, and along the California coast. There were not as many newspaper stories in 1981 as there had been when the second energy crisis began in 1979 (see figure 4.1), but energy was certainly a major issue that year.

The 1990 Field survey was conducted immediately after Iraq invaded Kuwait and President Bush ordered troops to Saudi Arabia to protect it from invasion, but long before there was any talk of sending U.S. forces to liberate Kuwait.[5] The Iraqi invasion had sparked an immediate surge in gas prices, but the price hikes were not nearly as sharp as the jumps in 1973 and 1979. Moreover, no gas lines ever developed. Energy issues naturally returned to the news immediately following the invasion, but the preceding months had seen relatively few stories about petroleum or anything else related to energy. In short, the 1990 survey was conducted at a time when a good deal of attention was being given to energy issues, but unlike the 1981 survey, the media coverage had been going on for only a few weeks.

The 1998 survey was conducted during a period in which energy issues rarely made the news. There was no energy crisis, there were no gas lines, and there was no war in the Middle East. To the contrary, oil prices were at an all-time low (adjusted for inflation) and were steadily falling. The economy was booming and there was not even a hint of any foreign entanglement related to petroleum.

In broad outline, our three surveys give us snapshots of public opinion over a seventeen-year period during which energy issues were fading from the public spotlight. This is the curious opposite of the situation that public opinion researchers usually study—the rise of new issues. These data, therefore, give us an opportunity to take a look at how the public thinks about issues that get less and less attention over the years.

ON LEARNING POLITICAL OPINIONS AND FORGETTING THEM

Opinions do not magically appear in people's heads. In some fashion, people must acquire them. To some extent, people learn their opinions from their parents, friends, and coworkers.[6] But this explanation begs the question, in that it presents something of a chicken-and-egg quandary: which came first? If children learn their opinions from their parents and later from their friends and coworkers, where did the parents, friends, and coworkers learn their opinions? In addition, how do people form opinions on entirely new issues—issues on which parents, friends, and coworkers have no views?

Philip Converse offers an answer to these questions in his classic 1964 paper "The Nature of Belief Systems in Mass Publics."[7] He argues that political leaders construct liberal, conservative, and other ideologies—which he calls "belief systems"—and teach them to the rest of the public. To understand how ideologies develop, Converse says, one has to recognize two basic points:

> First, the shaping of belief systems of any range into apparently logical wholes that are credible to large numbers of people is an act of creative synthesis characteristic of only a minuscule proportion of any population. Second, to the extent that multiple idea-elements of a belief system are socially diffused from such creative sources, they tend to be diffused in "packages," which consumers come to see as "natural" wholes, for they are presented in such terms ("If you believe this, then you will also believe that, for it follows in such-and-such ways").[8]

In simpler terms, Converse claims that people do not work out ideologies on their own; rather, ideologies are learned. Although people may add their own personal touches to the packages of opinions they learn, only a handful of the most influential elites make basic changes in ideological frameworks, and these elites guide the public in understanding how new issues fit into the established ideologies. For the rest of the population, ideologies are institutionalized sets of beliefs that are "out there" to be learned. Some people learn them quite well (and know, for example, that conservatives believe the government should not ban offshore oil drilling because the government should leave such decisions up to the free market); others learn only bits and pieces of them (and may know, for example, that conservatives tend to favor offshore oil drilling, but not exactly why); still others learn nothing at all.

Building on the research of Philip Converse and a number of psychologists, John Zaller fills in the details of Converse's answer by proposing a model to explain how people form opinions and answer survey questions regarding policy issues.[9] Zaller begins by assuming that everyone has certain basic values—or "predispositions," as he labels them. He distinguishes predispositions from the arguments or other information contained in the mass media, messages he labels "considerations." Beyond that, Zaller offers four basic propositions or axioms:

- The greater a person's level of cognitive engagement with an issue, the more likely he or she is to be exposed to and comprehend—in a word, to receive—political messages concerning the issue.
- People tend to resist arguments that are inconsistent with their political predispositions, but they do so only to the extent that they possess the contextual information necessary to perceive a relationship between the message and their predispositions.
- The more recently a consideration has been called to mind or thought about, the less time it takes to retrieve that consideration or related considerations from memory and bring them to the top of the head for use.
- Individuals answer survey questions by averaging across the considerations that are immediately salient or accessible to them.[10]

In other words, exposure to political news increases with political awareness and knowledge. For those who have been exposed, acceptance of messages increases with knowledge and awareness, *if* the messages agree with the person's predispositions. In contrast, among those who have been exposed, acceptance of messages *decreases* with knowledge and awareness when the message disagrees with predispositions. When people are asked survey questions, they quickly think of a few arguments related to the question and then answer based on those considerations, rather than on any deeply held, permanent attitudes.

The core of Zaller's theory is that people do not carry fully formed opinions around in their heads all the time; instead, they make up their opinions from whatever considerations occur to them at the moment they are asked questions. Here political awareness and knowledge play a central role. People who do not pay much attention to politics and are poorly informed usually have only a few, random considerations to draw upon. In contrast, people who pay a good deal of attention to politics and are well informed tend to have a good many considerations ready to hand. Moreover, their considerations generally match their basic values or predispositions because they critically evaluate news stories when they hear them. So, for example, well-informed liberals are likely to accept liberal claims made by the news media, but to reject conservative claims.

To describe these claims in concrete terms, we can think about the differences between typical high school dropouts who do not care about politics, and college graduates who do care. The college graduates will be far more likely than the high school dropouts to notice and register any given news item about, say, global warming. After all, they care more about politics, so they will read newspapers and pay attention to television news. Once having heard news items, however, the college graduates will be better prepared than the high school dropouts to decide whether they agree with the claims made in the news stories. Most pro-development readers will have read other articles about global warming, and many will have formed opinions that scientists who claim that global warming is occurring should not be taken seriously. A pro-environmental college-educated reader will also have an opinion, but he or she will have a predisposition to believe warnings about global warming. In both cases, the latest news item is typically just one more in a series of news items that the interested, college-educated person has read. In contrast, the high school dropouts who rarely pay much attention to politics will be much less likely to read or hear a news story about global warming. However, if such a person does read an article about global warming, then he or she will be likely to accept its claims uncritically, because of a lack of any other information with which to evaluate them.

The four panels in figure 5.1 illustrate a hypothetical relationship between political awareness and acceptance of the claims made in a pro-environmental newspaper article. The first panel shows that as awareness increases, people become more likely to read or hear the article. The second panel shows that as awareness increases, liberals who receive the message become increasingly likely to agree with the claims made in the newspaper article because the claims are consistent with their prior beliefs. The third panel shows that as awareness increases, conservatives who receive the message become increasingly likely to reject the article because its claims are contrary to their prior beliefs. The final panel shows the likely opinions of both liberals and conservatives when one considers both the likelihood of receiving the message and the likelihood of accepting it. As awareness increases, liberals become more likely to accept the message, as one might expect. Conservatives, however, initially become more likely to accept the message (because it is one of the few messages they receive) and then become less likely to accept it (because they realize that it contradicts their beliefs). The characteristic curve predicted by the model for this group is an inverted U, or ∩-shape.

The mathematics underlying this curious relationship are shown in table 5.1. The top row shows the likelihood of conservatives' receiving a message rising

Figure 5.1 Hypothetical Reception and Awareness Curves for a Liberal Message

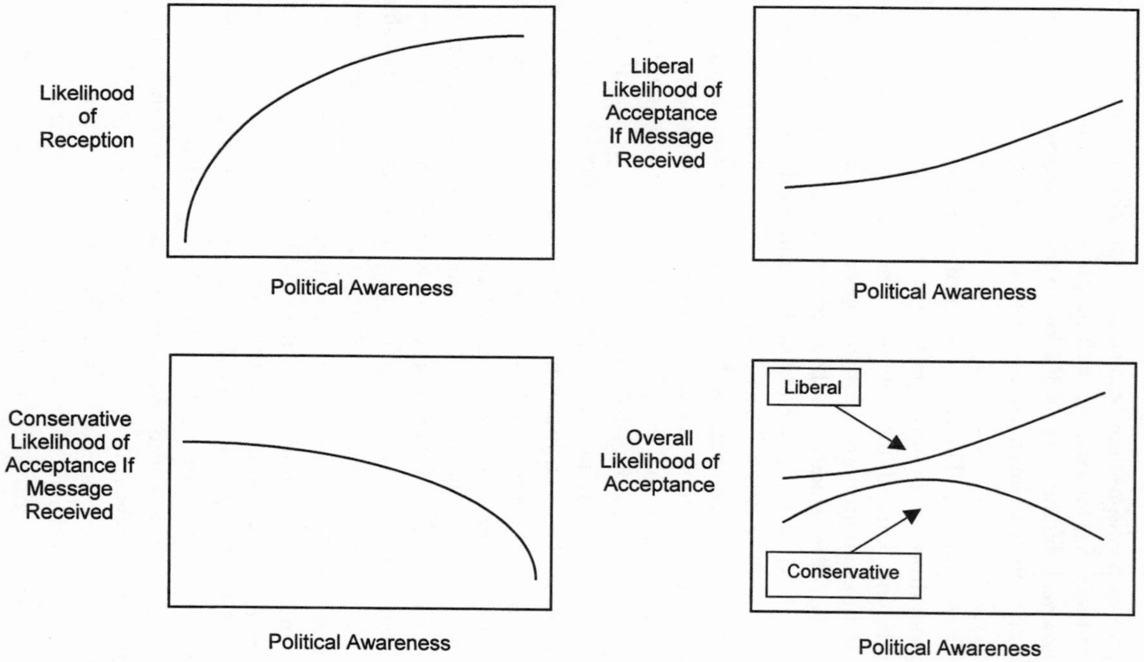

Table 5.1   Attitude Change in Response to a Hypothetical Controversial Message

Model: Prob (Change) = Prob (Reception) × Prob (Acceptance | Reception)

	Level of Awareness		
	Low	*Middle*	*High*
Prob (Reception)	.10	.50	.90
Prob (Accept | Reception)	.90	.50	.10
Change (Reception × Acceptance)	.09	.25	.09

Explanation:

Top row: As Awareness increases, the probability of receiving a message increases from .10 to .90.

Middle row: If the message is received, as Awareness increases, the probability of accepting or agreeing with a message decreases from .90 to .10.

Bottom row: The probability of attitude change is the product of the top and middle rows. That is, the probability of changing one's attitude is the probability of receiving the message multiplied by the probability of accepting it if it is received.

Source: John R. Zaller, *The Nature and Origins of Mass Opinion* (Cambridge: Cambridge University Press, 1992), 122–23.

from 10 percent to 90 percent as political awareness rises. The middle row shows the likelihood of their accepting the claims falling from 90 percent to 10 percent as awareness rises and the conservatives increasingly recognize that the liberal claim is inconsistent with their views. The bottom row shows the effect of the two forces together, calculated by multiplying the entries in the first and second rows. Conservatives at moderate levels of awareness are more likely to accept the message than either well or poorly informed conservatives.

Zaller's "Receive-Accept-Sample," or RAS, model helps explain a wide variety of phenomena, from unreliability in answers to public opinion questions to voting choices in congressional elections. For our purposes, however, the RAS model is most useful for explaining the differences between well-informed and poorly informed people and explaining how opinions toward energy issues have changed over time. Moreover, it serves as the foundation for the explanation of public opinion on potentially risky environmental technologies, which we will discuss at the end of the chapter.

Now that the theoretical stage is set, let us turn to an examination of public opinion toward energy issues.

THE CAUSES AND CORRELATES OF OPINION ON ENERGY ISSUES

To understand people's opinions on energy policy and other environmental issues, it helps to put those opinions into the larger context of what the public thinks about a wide range of issues. We begin our search for the causes of preferences on energy issues, therefore, by turning to past studies of public opinion and looking at the distinction between economic and social issues.

For many years, scholars have known that one of the best ways to describe public opinion on domestic issues is to divide them into two broad clusters—economic issues and social issues. Economic issues are those having to do with the distribution of wealth in society—taxes, welfare, Medicare, Social Security, most regulation of business and unions, and similar issues. Social issues have their bases in morals and value judgments—civil rights, women's equality, free speech, abortion and birth control, pornography, lifestyles, sexual preferences, and related issues. Richard Scammon and Ben Wattenberg originally popularized this distinction in their analysis of the 1968 presidential election, but since then it has come into wide use.[11]

For most of American history, the central differences between the two major, competing political parties have been over economic issues. Since the election of President Franklin Roosevelt and the beginning of the New Deal, the contest between the Democrats and Republicans could usually be described (although perhaps somewhat simplistically) as that of the party of labor versus the party of management, or the party of the poor versus the party of the rich. In the 1960s, however, new issues arose that could not easily be fit into economic terms—civil rights, women's equality, and others. These issues were not entirely independent of the traditional economic issues. People who were liberal on economic issues tended to be liberal in social matters as well. To put it in more concrete terms, people who favored increasing the minimum wage were also likely to favor civil rights for blacks. Nevertheless, these new social issues tended to stand out from the old economic issues. Support for civil rights, women's equality, free speech, and for other socially "liberal" causes was widespread among well-educated, upper-income people who held conservative views on issues such as income taxes. The working class, however, was not especially fond of these radical new ideas. The result was that public opinion researchers found it useful to group issues into social and economic categories when they explained them.

The distinction between economic and social issues is useful not only because opinions on the two types of issues cluster together separately, but also because opinions on economic and social issues seem to have different causes. Economic issues usually vary with family income, while opinions on social issues usually vary with education and age. This generalization offers us a starting point for our search.

Income: Energy as an Economic Issue

Several explanations have been proposed for why we should expect people's opinions on energy issues to depend on their wealth or their family incomes.

First, energy issues are economic issues. Energy production is a major sector of the economy. Most environmental policy disputes over energy involve government regulations on private businesses, which cost the businesses money. Sometimes the regulations cost people their jobs; at other times they create new jobs. From controls on pollution and toxic wastes to protections for endangered species, environmental laws and regulations have substantial economic effects.

Since the early days of the oil industry, people have fought over taxes and other revenues from the industry. Laws regulating the industry have sought a range of economic goals—from protecting U.S. jobs and the profitability of U.S. oil firms in the 1950s to cushioning the American public from the impact of the oil shocks in the 1970s. Some laws were even explicitly designed to redistribute wealth from high-income energy users to low-income users. During the 1980s, when energy policy was last on the front burner of American politics, the Reagan administration pushed aggressively for more oil development and the construction of more nuclear power plants. President Reagan, Interior Secretary James Watt, and other administration spokespeople couched their arguments in explicit economic terms. They declared that the best energy policy was for the government to relax regulations that limited energy production and leave our energy fate up to the decisions of free markets. More generally, supporters of energy development usually appeal to the public in terms of economic benefits. Oil and nuclear power provide jobs, taxes to local communities in which the developments are located (as well as to the state and federal governments), and cheaper energy for economic growth.[12]

Although energy is an economic issue, how this plays out in public opinion is not entirely clear. The basic pattern one finds among economic issues is that people tend to prefer policies that will economically benefit themselves or people like them. Most obviously, the wealthy tend to prefer lower taxes (especially on themselves) and cuts in government services (especially services going to the poor). The poor, in contrast, generally prefer higher taxes (especially when paid by the rich) and more government services (especially those targeting the poor). Similarly, other people who benefit from a government program are generally more enthusiastic about the program than those who do not benefit. Blacks, for instance, favor affirmative action programs and programs that aid low-income groups because blacks are disproportionately working class. Students are especially fond of student-aid programs. Farmers may not like big government, but the agricultural price supports that big government once gave them were generally appreciated. In other words, opinions on economic issues often depend on economic self-interest—hardly a big surprise.[13]

The biggest direct beneficiaries of energy developments tend to be upper-income people. Jobs in both the oil and nuclear industries pay well above median salaries. The stockholders of energy firms, who benefit from industry expansion, are like stockholders in other sectors of the economy: they tend to be wealthy. As a consequence, we might expect high-income people to favor energy development more strongly than do those with lower incomes. This pattern is not entirely obvious, however, because all of us are indirect beneficiaries of energy production. Low energy prices, after all, benefit everyone in the economy.

A second line of argument about income, and social class more broadly, is that upper-income people have filled most of their material needs and can therefore pay more attention to quality-of-life issues, such as the state of the environment. They have enough disposable income to pay for clean air, clean water, and clean energy without worrying about the cost. Those with lower incomes face more painful trade-offs. The higher costs of low-pollution, high-gas-mileage cars may mean no cars for the poor. These arguments lead one to expect that as income increases, people will lean more strongly to the pro-environment side—the opposite of the first prediction. Although this second argument runs contrary to the conventional interpretation of economics issues, it has received a good deal of support from researchers on public opinion toward environmental issues.[14]

The data support neither of the above arguments. In the cases of offshore oil development and drilling in government parks and forest reserves, no relationships exist. The correlations are all tiny and statistically insignificant. In order to avoid burying readers in tables and figures, the data are not shown, but interested readers can see them on the Web.[15]

The results for the nuclear power question differ. In figure 5.2 we see that the poor are significantly less likely to support nuclear power than are middle and upper-income groups. In this case, support for nuclear power looks like a con-

Figure 5.2   Support for Nuclear Power by Income

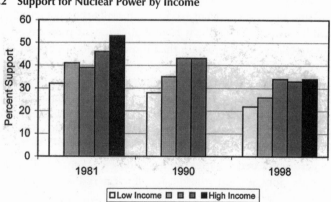

ventional economic issue: the wealthy are pro-development and the poor are pro-environment.

As for whether our energy issues look like economic issues, the evidence is mixed, but points toward a negative answer. The two questions about oil development certainly do not show the typical relationships found among other economic issues. The nuclear-power question suggests an economic issue, but—as Dunlap and Van Liere have shown—the relationships between income and most questions about environmental issues look like the oil questions rather than like the nuclear-power question.[16] That is, income correlates with opinions on a few environmental issues, but in most cases, income and opinions are unrelated. Moreover, there is some question about whether nuclear power is a typical environmental issue. One study, for example, argues that it is atypical because it is far more ideological and partisan than other environmental issues.[17] So although some might say that environmental issues are conventional economic issues from an analytic point of view, they do not appear that way in the public's mind.

Education and Age: Energy as a Social Issue

The patterns of opinion on social issues are more complicated than those on economic issues. Tolerance lies at the heart of many social issues—tolerance for people with different religions, for people of different races or ethnicity, and for people whose moral judgments differ from one's own. In general, the most tolerant people—and consequently the most liberal on social issues—are the well educated and the young.

Most researchers argue that education broadens people's perspectives, brings them into contact with people who are not like themselves, and shows them that they have nothing to fear from people who are different. The more people learn about another group—blacks, Latinos, the disabled, gays and lesbians, communists and right-wing militia members alike—the more people tolerate them. Education does not make people agree with other groups, but it does make them more likely to believe that everyone ought to be treated equally and to be protected from discrimination. Researchers also argue that the best educated are best able to learn society's prevailing norms and values, and that they are exposed to more years (at college) of intensive socialization. Because the prevailing norms in the United States are that people should be tolerant, the best educated learn to be the most tolerant. Exactly why education leads to tolerance is not important here; we merely need to observe that two by-products of education are tolerance and liberal views on social issues.[18]

Age is also associated with opinions on social issues, but not because age causes tolerance. Instead, age is related to social liberalism because age reflects the generations in which people grew up and because it reflects the rising educational levels of each new generation. Throughout the century, American society has been moving in a more tolerant direction. Ideas such as racial and gender equality which were once considered radical are now widely accepted. Many of these ideas have been passed into law. Not only is blatant racial bigotry scorned by most people, but racial discrimination is illegal in many circumstances. The same applies to discrimination against women, people with unpopular religions or opinions, the disabled, older workers, and in some places even gays and lesbians. Decade by decade, tolerance has been spreading. And decade by decade, the young have been socialized to accept the more tolerant and socially liberal views of society. Children and teenagers growing up in the 1920s, for example, learned very different values than children and teenagers growing up in the 1970s. As a result, the changing social norms and laws have made each generation more liberal on most social issues than were previous generations.[19]

In the past, some scholars speculated that life-cycle effects (discussed in chapter 3) rather than generational changes might cause the relationship between social liberalism and age. They suggested, as we have seen, that as people grow older, they naturally change their views on a range of issues. The young are naturally tolerant and liberal on many social issues, so the argument goes, because they are more flexible and more willing to challenge the status quo than are older people. As they age, they lose their flexibility, become more attached to the status quo, and become more conservative. Indeed, this argument is so firmly entrenched that many people accept it as folk wisdom.[20] As we saw, however, the life-cycle argument may sound attractive, but the evidence does not support it.

Another reason why age is strongly correlated with liberal social views is that the young are, on average, much better educated than the old.[21] In fact, throughout this century, every generation has been better educated than the previous generation. People born in the 1930s, for instance, spent fewer years in school than those who were born in the 1940s, who in turn spent fewer years in school than those born in the 1950s, and so on. Elementary-school dropouts were common early in the century, but by the 1960s and 1970s, they had all but disappeared—except in retirement communities. So when one compares people born in, say, the 1920s to people born in the 1970s, one is comparing groups with substantially different levels of education. The better education of the young, of course, helps explain why the young are more tolerant and liberal on social issues than the old.

To summarize, we have two patterns in public opinion data. Opinion on economic issues normally varies with income and other variables that reflect clear economic interests. Opinion on social issues normally varies with education, age, and variables that reflect tolerance.

The surprise is that opinions on most environmental issues look like opinions on social issues, not economic issues—and the energy issues we will examine in this chapter generally fit that pattern. Although previous studies have not put environmental issues in the context of economic and social issues, the patterns they have found fit that distinction. Mayer, for example, finds that spending for environmental protection and educational purposes were the only two "economic issues" that were related to age or education.[22] Ronald Inglehart declares that "postmaterialist issues" (his term for social issues) include "environmentalism, the women's movement, unilateral disarmament, [and] opposition to nuclear power."[23] A number of other studies have produced similar findings about a wide range of environmental issues.[24] Even though environmental issues do not depend on tolerance or moral or religious values in the same obvious way that other social issues do, when we examine public opinion toward these energy issues in detail, we see that the relationships look like typical social issues. When we begin thinking of environmental issues as social issues, we can begin to make some progress in explaining them.

We begin our examination of environmental issues as social issues with education. On economic issues the well educated tend to be somewhat conservative because they usually have high income jobs, and the poorly educated tend to be somewhat liberal because they tend to have low-income jobs. On social issues, however, the relationship reverses—the well educated tend to be liberal, and those who never graduated from high school tend to be conservative. In the two questions about oil development in figure 5.3, we see the relationship of a typical social issue. For instance, in 1998, 35 percent of those who never completed high school favored more offshore oil drilling, but only 22 percent of those with education beyond college favored more drilling. Similarly, 41 percent of high school dropouts favored more drilling in government forest reserves and parklands, but only 27 percent of respondents with postgraduate educations took that stand. That is, the best-educated people in our sample are substantially more likely than the least well educated to oppose oil development of any kind. That is exactly the sort of relationship one expects to see with a social issue. Moreover, this relationship is much stronger than any of the relationships between income and opinion about oil drilling.

The relationship between education and opinion on nuclear power is more complicated. In 1990 we see that the least well educated are the most strongly opposed to construction of more nuclear power plants. Support for nuclear power increases with each educational group until we reach the college graduates, who are the most supportive, but then opposition declines again in the best-educated group—people with postgraduate degrees. This is the opposite of the pattern found in opinions about offshore oil development, and it does not support the claim that people respond to nuclear power in the same way they respond to typical social issues. The results for 1981 and 1998 are less clear. The

Figure 5.3   Support for Energy by Education

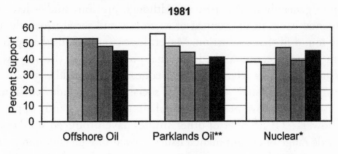

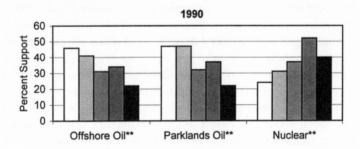

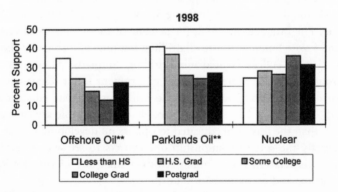

*p < .10; **p < .05

well educated show a slightly higher level of support for nuclear power, but the differences are not statistically significant. That is, they could simply be the result of sampling error. In any case, what we do not see is a pattern typical of social issues—a pattern of the best educated taking the most pro-environmental stands.

This pattern of mixed findings about the relationship between education and attitudes toward nuclear power matches the findings of previous studies of the question. Most studies have found that opposition to nuclear power increases with education—as if nuclear power were a social issue.[25] Other studies have found that opposition decreases with education.[26] Still others find that education is unrelated to opinions about nuclear power.[27] These studies were conducted using different methods (public opinion surveys and aggregate analysis of election data) and in different contexts (in samples drawn from communities near proposed nuclear power plant sites, and in general population samples), which may explain the discrepancies in findings. Still, collectively these studies suggest that the relationship of education to opinions on nuclear power may be more complicated than the relationships of education to other environmental issues. It may be, as David Webber suggests, that nuclear power is not a typical environmental issue.[28]

Exactly what is behind the education-environmental opinion relationship is not clear. With most social issues, researchers argue either that knowledge about other groups breeds tolerance, which underlies liberal social views, or that the well educated are good at learning society's norms and have been socialized for a longer time. Are political knowledge and learning ability (practically indistinguishable with survey data) causes of environmentalist views? We explore this possibility by examining the relationship of political knowledge to opinions on energy sources.

Research on the relationship between knowledge and environmental attitudes falls into two groups. Researchers in the first group have explored people's perceptions of environmental risks—for example, the risks stemming from nuclear power plants. The findings here have been mixed. Some studies suggest that risk perceptions depend on information.[29] Most studies, however, point to the conclusion that deeply held values, cultural orientations, and other factors govern risk perceptions, and that knowledge about potential hazards has little or no influence.[30] In fact, Wildavsky flatly declares, "knowledge of actual dangers makes no difference whatsoever" in risk perceptions.[31]

Researchers in the second group have investigated the relationship between knowledge and broad environmental attitudes, rather than just focusing on risk perception. These investigators have generally found that people who are more knowledgeable are more likely to hold pro-environmental views.[32] The weakness of these studies—a weakness shared by the risk-perception studies—is that they mostly use measures of knowledge about environmental issues, rather than general knowledge. This leaves open the question: Do knowledgeable people learn about environmental issues and become environmentalists, or do people who are very concerned with the environment learn about environmental issues? In other words, does knowledge cause attitudes, or is the causal relationship reversed, with attitudes causing knowledge?[33]

In order to resolve this problem, we use a measure of general political knowledge—the same knowledge index used in chapter 4. If there is a relationship between this general knowledge index and environmentalist views, presumably knowledge (or superior learning ability) is causing opinions. After all, it is not very plausible to argue the reverse—that people who are especially interested in the environment will dash out and start learning about *all* political issues. Broad political knowledge is a general characteristic (or accomplishment, if you will), not something characteristic only of people who care about the environment.

The first step in testing our hypothesis is to look at the simple relationship between knowledge and environmental attitudes. The relationship looks almost exactly like the one shown in figure 5.3 (and therefore I omit the data). The higher one scores on the knowledge index, the less likely one is to support either offshore or parklands oil drilling, but the more likely one is to support nuclear power.

To explore the relationship further, table 5.2 presents three simple regression equations using education and political knowledge as the independent variables to explain attitudes toward offshore oil drilling, parklands drilling, and nuclear power. The results reveal that political knowledge is at the core of the relationship. In the equation explaining attitudes toward offshore oil development, the

Table 5.2   Regression Models of the Effects of Education and Knowledge on Opinions, 1998

	Offshore Oil			Parklands Oil	
	Coefficient	p-value		Coefficient	p-value
Education	−0.08	0.094	Education	−0.06	0.247
Knowledge	−0.07	0.026	Knowledge	−0.18	0.000
Intercept	2.47	0.001	Intercept	3.07	0.000
N	808		N	808	
Adjusted R²	0.01		Adjusted R²	0.03	

	Nuclear Power	
	Coefficient	p-value
Education	0.02	0.630
Knowledge	0.07	0.071
Intercept	1.89	0.000
N	808	
Adjusted R²	0.004	

Note: **Bold** coefficients are at least of borderline statistical significance ($p < .10$).

two variables have roughly equal impacts, with knowledge being statistically significant at the usual .05 level and education attaining borderline significance. In the other two equations, however, knowledge has a far larger impact than education, which fails even to come close to statistical significance in either equation. In short, differences in general political knowledge seem to explain the relationship between education and environmental attitudes.

There is more to be learned about the role of knowledge, but it will have to wait until more of the foundation is laid out. We will return to investigate the role of knowledge in more detail in the final section of this chapter, where we will see how knowledge interacts with cultural values to explain opinions.

Turning to the relationship between opinions on energy issues and age in figure 5.4, we see basically what we would expect of a social issue. Support for both oil drilling and nuclear power generally increases with age. The oldest people in our samples are the strongest supporters of all three types of energy development, while the young are usually the most pro-environmental. These results match those of a host of other studies showing that age predicts environmental attitudes as well as—or better than—any other variable.[34]

A closer look at the data in figure 5.4 shows that there is more to the story than the idea that environmentalism is strongest among the young. The strength of the relationship between age and support for energy production weakens across our three surveys from 1981 to 1998. The relationship was strongest in 1981, at the end of the second energy crisis. In that year, the differences between the young and the old were more than 20 percent. By 1990, the differences on offshore oil development between the young and the old had diminished. Those over sixty were now only about 11 percent more likely to support offshore drilling than the youngest group. By 1998, the differences between the young and old had diminished on all three energy-production questions. Those over sixty were only about 5 percent more likely to support offshore oil development and 7 percent more likely to support nuclear power than were the young. In addition, there were no longer any real differences among those under sixty on either of those questions. On the matter of drilling in government park lands and forest reserves, the young and old supported the idea at similar levels, while those in between were more environmental. Moreover, the difference between the most and least pro-oil age groups was less than 10 percent. In short, from 1981 to 1998, the relationship between age and opinions on energy production fell apart.

What happened? Zaller's RAS model offers an explanation: the decline in news coverage of energy issues. In 1981, the United States was just coming out of an energy crisis. For years, the public had been bombarded with news about oil issues and nuclear power (see figure 4.1). People had been told over and over by political leaders and journalists that environmentalists opposed oil drilling

Figure 5.4   Energy Opinions by Age

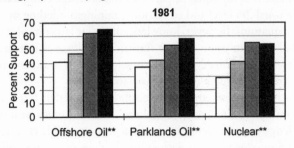

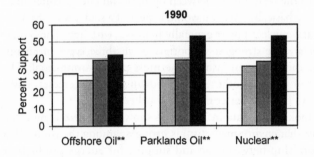

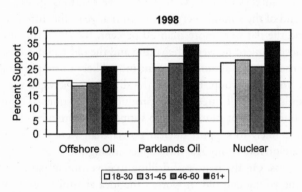

*p < .10; **p < .05

and nuclear power, while business and pro-development leaders supported them. They learned those lessons in the 1970s, but without the stream of reminders, people began to forget. By the late 1990s, of course, a new generation had grown to adulthood without the threat of oil shortages. These people never had much opportunity to learn what they should think about offshore oil drilling. This is where Zaller's "accessibility axiom," quoted above, comes into play:

"The more recently a consideration has been called to mind or thought about, the less time it takes to retrieve that consideration or related considerations from memory and bring them to the top of the head for use." To this comment, Zaller added: "Conversely, the longer it has been since a consideration or related idea has been activated, the less likely it is to be accessible at the top of the head; in the limit, a long unused set of considerations may be completely inaccessible, which is to say, forgotten."[35]

In other words, the RAS model implies that when the news media no longer cover a political issue, people will forget (or never have the opportunity to learn) which political positions should be associated with which sets of values or ideologies. In Converse's words, people will fail to learn "what goes with what." Young environmentalists, for example, will not realize that they should consider offshore oil drilling to be bad, while their pro-development opponents will not learn why it is good. The RAS model's explanation for the weakening of the relationship between age and opinions on energy issues, then, is that news-media coverage of energy issues declined, reducing the number of considerations people had to call upon when asked questions about energy policy. The lack of knowledge about energy issues made pro-environment and pro-development people's opinions more similar, because few people knew enough about the issues to recognize the pro-environment and pro-development sides. In terms of the last panel in figure 5.1, America moved from high awareness toward low awareness (that is, from right to left), and the differences among groups diminished.

We will continue to explore changes in opinion over time in the rest of this chapter; however, at this point we can see that change over time is more than a matter of change in aggregate opinion. Chapter 3 described overall trends in opinion. Here we see that those aggregate trends masked other trends among subgroups. Different groups changed in different ways as energy issues faded from the daily news. This pattern will come up several more times in this chapter.

Energy, Race, Ethnicity, and Gender

Another set of characteristics that are often related to public opinion are race and ethnicity. On most economic issues, whites tend to take more conservative stands than nonwhites, because as a group whites are higher up the economic ladder. On social issues, researchers have found more of a mixed bag. On any issue related to civil rights or questions of racial and ethnic tolerance, whites are more conservative than nonwhites. However, on social issues unrelated to civil rights—abortion or gay and lesbian rights, for example—nonwhites are often more conservative than whites.[36] What we should expect, therefore, is not entirely clear.

In the case of the two questions about oil development, the data reveal changing patterns and weak associations.[37] African Americans were the most likely to support both types of oil development in 1990, but they were the least likely to support offshore drilling in 1998. In 1981, the differences were too small to be significant. The support levels of Asians and Latinos also move around a good deal. Only in the case of attitudes toward nuclear power does a consistent pattern emerge—whites are always more pronuclear than any other group.

Differences also appear over time. In response to the oil drilling questions, there were no significant differences in 1981. By 1998, however, whites had become generally more opposed to drilling than nonwhites. Their attitudes changed more quickly over time. Responses to the nuclear power questions were quite different. In all three surveys from 1981 through 1998, whites were consistently more supportive of nuclear power than were nonwhites. The overall level of support for nuclear power declined across the three surveys, but the gap between groups did not.

Moving beyond the details, one might ask, how should we characterize racial and ethnic attitudes toward energy issues? There have been a number of studies of this question, all of which have suffered from two shortcomings. First, they have focused exclusively on white-black differences. Latinos and Asians have been ignored. Second, they have relied heavily on questions about environmental concern and have largely ignored specific environmental issues. Strictly speaking, there is nothing wrong with looking only at environmental concern, but it may encourage readers to go beyond the data and infer that if there are no differences in how important environmental issues are, there must be no differences in policy views either, which would be a mistake.

Some early studies of the question suggested that blacks were less concerned than whites with environmental issues.[38] Other, more recent and more comprehensive studies contradict the early studies.[39] They find that blacks are less likely to become actively involved in environmental politics, but are no less concerned than whites about environmental threats. The data here go even further, suggesting that the answer depends both on the time and on the specific environmental question being asked. In the case of oil development, whites sometimes lean in favor of development, sometimes lean in favor of the environment, and sometimes are the same as other racial and ethnic groups. Only in the case of nuclear power is there a consistent pattern: blacks are always more pro-environment. As we shall see below in our multivariate analyses, education and other factors explain some of these differences. Still, we can reasonably conclude that blacks may be more or less pro-environmental than whites, *depending on the issue.* The matter is not as simple as previous studies have suggested.

Since a gender gap appeared in voting choices at the end of the 1970s, a grow-ing amount of research has sought to explain that gap by examining the differ-ences between men's and women's opinions on a variety of policy issues. In some areas, such as the use of force and violence, studies have consistently found differences between women and men. In other areas, such as women's rights, few gender differences have been found. In the area of environmental politics, however, the findings have been mixed. Virtually all studies agree that men sup-port nuclear power far more strongly than do women; beyond that, however, gender gaps appear on some environmental issues, but not others.[40] The appar-ent key to explaining these results is that women take stronger environmental positions when they perceive some type of direct threat to people (for example, from pesticide use or nuclear power), but not otherwise (for example, in spend-ing on parks). If we assume that people believe that nuclear power poses a serious threat to human beings and that oil drilling poses little or no threat, we can explain our findings accordingly.

The data show that men always favor development more than women do, although the differences are not always large or statistically significant. The key difference is between attitudes toward oil and toward nuclear power. The find-ings are mixed on oil drilling. Men were significantly more likely to support drilling in three of the six oil questions—in 1981 on offshore oil drilling, and in 1998 on both kinds of oil drilling. On nuclear power questions, however, men consistently held more favorable views than women, and they did so by substantial margins (ranging from 13 to 18 percent). In sum, we see a large gender gap on nuclear power and a smaller, inconsistent gap on oil drilling.

Where One Lives

Does it matter how close one lives to the coast, to an offshore drilling rig, or to a nuclear power plant? Most observers assume it does. The word NIMBY, or "not in my backyard," has come into common usage to describe efforts by local activists to resist proposed developments in their neighborhoods. Similarly, land use planners speak of LULUs or "locally unwanted land uses." Given that both nuclear power and offshore oil development have been the targets of mass pro-tests by residents near proposed developments, one might reasonably infer that the NIMBY syndrome is alive and well.[41] Still, some nuclear and oil industry defenders claim that the protesters are small, unrepresentative groups of vocal extremists who represent no one but themselves. Surprisingly, however, public opinion researchers have not extensively investigated whether how close one lives to a nuclear power plant or offshore oil facility is related to views about them.

The lack of studies of the effect of proximity no doubt stems from the diffi-culty of investigating the problem. National public opinion surveys—the most common type of survey used by academics—do not include enough residents living close to nuclear power plants or offshore oil wells to study. Consequently, these studies can only ask abstract questions—about, for example, whether people favor nuclear power in general, and whether they would favor the construction of a power plant within five miles of their homes. In contrast, studies of particular communities near nuclear power plants or oil drilling operations can ask detailed questions about real or proposed facilities in the area, but they suffer from the possibility that the communities are atypical and that the results cannot be generalized to the rest of the United States. In other words, although this approach gives us insight into questions about proximity, it also carries the potential of confusing distance from a facility with other characteristics of the communities.

Studies of the effects of proximity have, in fact, produced somewhat mixed results. In general, studies based on national public opinion surveys have found that people are much more likely to say they favor the construction of nuclear power plants if the location is not stated in the question than if told the plants will be built in the area (for example, within five miles).[42] Local studies, however, have not yielded such systematic opposition. In one study, people living close to a nuclear power plant were more likely than those living farther away to oppose it when it was about to open, but other studies indicate that the experience of living near a nuclear power plant seems to cause people to look upon them more favorably.[43] Questions about proximity remain.[44]

In order to discover whether people living near offshore oil drilling sites or nuclear power plants differ from people living elsewhere in the state, several measures of proximity to those developments were examined. The first measure of proximity to be examined was whether people resided in a county with either offshore oil development or a nuclear power plant. To be more precise—in the case of offshore oil drilling, the opinions of people living in Santa Barbara or Ventura Counties were compared to the opinions of people living elsewhere in the state. Santa Barbara and Ventura stand out because they were the two counties directly affected by the 1969 Santa Barbara Channel oil spill and because local residents held massive, anti-oil protests. The other oil-producing counties (which were examined separately) are Los Angeles and Orange Counties, both of which include large population centers many miles from the ocean, unlike Santa Barbara and Ventura Counties. The counties with nuclear power plants are Humboldt, Sacramento, San Diego, and San Luis Obispo.

The data showed only small differences between people living in oil counties and those living elsewhere. In 1981, people living in oil-producing counties were a few percent more likely to oppose offshore oil drilling, and in 1998, they were a few percent more likely to favor offshore drilling. Neither of these differences was statistically significant.[45] A similar analysis was conducted for people who did or did not reside in counties with nuclear power plants in 1990 and 1998.[46] In this case the differences were both larger and statistically significant, but they indicated opposite results. In 1990, people living in counties with nuclear power plants were 11 percent less likely to favor nuclear power than people who lived elsewhere, but in 1998 exactly the opposite pattern prevailed. People who lived in counties with nuclear power plants were 11 percent more likely to favor nuclear power. In short, the data provide very little support for the claim that proximity matters with respect to people's opinions on either offshore oil development or nuclear power.

So as to be more confident about the finding that proximity to offshore oil development or to the coast makes no difference to people's attitudes, several different measures of proximity were tested to see whether any measure of proximity would work. Using both simple difference-of-means tests and multivariate models (discussed later), I first looked at differences between respondents living in coastal counties and noncoastal counties. Second, I separated the more rural/ suburban oil-producing, coastal counties (San Luis Obispo, Santa Barbara, and Ventura) from the more urban oil-producing counties (Los Angeles and Orange) to see if either of those two regions differed in some fashion. Third, using Zip codes, I looked at differences between people living within twenty miles of the coast and those living outside that coastal band. Finally, I looked at how those living within twenty miles of the coast in a county with offshore oil development differed from those living elsewhere, and how those living within twenty miles of a nuclear power plant differed from those living elsewhere. I performed these analyses in several data sets, including the three data sets used in this chapter. In *none* of these analyses did people living near offshore oil drilling oppose those developments more than did people living elsewhere in the state. In the case of nuclear power, the only survey showing that people living near a nuclear power plant oppose nuclear power is the 1981 Field survey mentioned above. In that case, however, a multivariate analysis (reported later in this chapter) showed that other factors explained the relationship. Living in a county with a nuclear facility had nothing to do with attitudes. In sum, no NIMBY effects appeared in any of these data.

This result may seem surprising, since local opposition to oil development and nuclear power has been very intense. Still, this finding is consistent with previous studies from England showing that people who live close to oil wells or nuclear power plants are somewhat more positively inclined toward them than those who live farther away.[47] Other possibilities exist, of course. It may be that if we were to look more closely so that we could measure proximity within a few miles, rather than by county or Zip code, we would find NIMBY effects. Alternatively, perhaps NIMBY effects are temporary and appear only briefly— from the time a new development is proposed until construction has been completed. Yet regardless of the truth of these possibilities, and contrary to the impression produced by news coverage, the NIMBY syndrome does not seem to be active in people living near oil or nuclear power facilities in California.

So far we have been discussing how people respond to questions that do not specify exactly where the new oil drilling will take place (other than along the coast) or the nuclear power plants will be constructed. Perhaps if a question specified proposed sites for new development, the answers might change. In other words, perhaps we should be careful about generalizing from an abstract question to any specific proposed energy development.

Luckily for this investigation, two surveys of Santa Barbara county residents, conducted in 1988 and 1990, helped to fill this gap by asking questions about support for "oil drilling off the Santa Barbara coast."[48] In those years, there were highly publicized proposals to drill additional offshore wells along the Santa Barbara coast. The first survey yielded 39 percent in favor of drilling and 54 percent opposed; the second yielded 37 percent in favor and 54 percent opposed. These numbers are quite close to the 34 percent in favor of additional drilling found in the statewide 1990 Field Poll question.[49] That is, Santa Barbara residents were—if anything—slightly more favorable to the idea of offshore oil drilling than were other Californians. These data point toward the same conclusion as our other data: proximity to the coast or potential drilling sites does not cause opposition to offshore oil development.

Together these data provide a strong case that proximity to existing offshore oil wells has essentially no impact on support for additional development. In fact, we could uncritically accept that conclusion, were it not for the results of a small, unrepresentative data set—a college classroom survey.

In March 1993, Sonia Garcia and I administered a survey to an introductory American politics class for freshmen and sophomores at the University of California, Santa Barbara.[50] Two hundred forty-three students returned the questionnaires. The course fulfilled several graduation requirements and consequently included a wide range of students, not just future political science majors. However, by no stretch of the imagination were the students a representative cross-section of UCSB students, let alone Santa Barbarans or Californians.

Two of the questions we asked the students address the effects of proximity to oil development:

- "Suppose new offshore oil development platforms were planned to be built in Santa Barbara County. Would you support or oppose their construction?"
- "Suppose new offshore oil development platforms were planned to be built in a remote area off the California coast. Would you support or oppose their construction?"

The results: 12 percent of the respondents supported Santa Barbara oil development, while 80 percent opposed it. But when the question asked about a remote area of the California coast, 20 percent supported development, while only 59 percent opposed it.[51] Although majorities of our student respondents opposed oil drilling in both locations, some of the students in our sample opposed local drilling and approved of it elsewhere. The results for the two questions differed enough to attain statistical significance.

The implication from the student survey is that we ought to be careful in accepting the conclusion that proximity makes little or no difference to opinions about offshore oil development. Proximity may make a difference, but perhaps only when the proposed development is—as the NIMBY acronym implies—literally in one's backyard. For many students, oil development is in their backyard. Students living in Isla Vista, which is located on bluffs overlooking the ocean, can easily see oil platforms. Because of natural seepage from the sea floor, unrelated to wells, in the area, tar regularly washes up on the beaches surrounding Isla Vista and the UCSB campus, and local residents can often smell oil. For most of Santa Barbara County, however, the oil development would not literally be in their backyards. For most of the county, oil production is something seen at a distance, not over the back fence.

In sum, because the student survey found a higher level of support for oil development in remote areas than in Santa Barbara, we cannot be sure that the conclusion about proximity making little or no difference is correct. Unfortunately, there have been no local or statewide surveys that would allow us to investigate the role of proximity in public reactions to proposed or existing oil developments. Therefore, all we can say is that the evidence strongly points to the conclusion that proximity to oil production has little or no effect on attitudes, but we cannot be absolutely certain.

The finding that proximity to the California coast is unrelated to attitudes toward oil development warrants special attention. In their perceptive study of the politics of offshore oil development, *Oil in Troubled Waters*, Freudenburg and Gramling open their third chapter by paraphrasing an unnamed senior official from the U.S. Minerals Management Service, the government agency in charge of offshore oil development: "'The problem, he said, was that the agency was only hearing from 'the activists who make a lot of noise,' but he saw them as a small and unrepresentative slice of the population. The 'silent majority' of northern California residents, he continued, were actually in favor of leasing."[52]

Based on their analysis of a series of in-depth interviews and their reading of public testimony at various hearings, Freudenburg and Gramling argued that the official was wrong. The public, they maintained, was strongly opposed to offshore oil drilling. Our statewide, public opinion survey data confirm Freudenburg and Gramling's assessment: A substantial majority of the public is strongly opposed to additional offshore oil drilling.

We can add to Freudenburg and Gramling's assessment by observing that not only is the prevailing opinion among coastal residents against further offshore oil drilling, but the prevailing opinion in the state as a whole is also against drilling. Although coastal residents probably differ from other Californians in the intensity of their views about offshore oil drilling, nothing in the data suggests that they differ from noncoastal residents in the thrust of their opinions about proposals for further offshore oil development. A substantial and growing majority opposes offshore oil drilling.

Another aspect of where one lives is the size of the community in which one resides. Does it make any difference whether a person lives in a rural or urban area? People living in rural and urban areas, after all, have quite different relationships with their environments.

Two reasons have been proposed to explain why urban residents should be more pro-environment than rural residents. First, rural dwellers generally have more utilitarian relationships with their environments than do urban dwellers. Either they or people they know are likely to work in farming, ranching, mining, or logging. They may, therefore, be more willing to accept environmental sacrifices in exchange for economic gains. Second, rural dwellers may live in less polluted environments and so may be less concerned with environmental problems. Smog, overcrowding, and many other environmental problems are more typical of big cities than the countryside, or so this argument goes.[53]

Previous studies have indicated that urban dwellers are generally more likely than rural dwellers to favor environmentalist stands, but the findings have not been consistent.[54] In some cases, the relationship fails to appear and in a few cases it is reversed. David Kowalewski points out that this pattern of mixed results may stem from several causes.[55] First, it is not yet clear whether the relationship should be conceptualized in rural-urban terms or occupational terms. That is, should we be looking at whether people work in farming, ranching, mining, or logging, rather than at rural-urban differences? Indeed, one study suggests that farmers are more environmentally inclined than nonfarmers in the same communities.[56] Second, the dependent variables often change from study to study. Although more studies look at environmental concern than at any other variable, there are studies looking at opinions on a range of policies, including conservation, recycling, and government spending on the environment. In one of the more telling studies, McEnvoy found that people living in urban areas cared more about air pollution than water pollution, while people living in rural areas cared more about water pollution.[57] Arguably, each group cares more about the worse problem in its own area. In other words, we may be seeing no more than self-interest in these findings. More generally, we have to be careful about overgeneralizing about environmental issues. Third, some studies showing rural-urban differences look only at simple relationships, while others look at the effects of rural-urban differences controlling for other variables in multivariate models. Given the differing approaches, we should not be surprised at inconsistent results. In sum, the question of whether rural-urban differences have an effect on opinions is not yet answered.

Whether we should expect to find rural-urban differences in attitudes on energy policy is not obvious. Neither nuclear power nor oil drilling is primarily a rural or urban industry. Nuclear power plants are located in rural areas (Humboldt and San Luis Obispo Counties) and near urban centers (Sacramento and San Diego). Similarly, both onshore and offshore oil drilling operations exist in rural areas (Santa Barbara County and the ranch lands east of it) and in urban areas (Long Beach and Los Angeles—where oil wells are hidden inside fake office buildings on Wilshire Boulevard). Consequently, neither the pollution problems nor the jobs associated with nuclear power and oil can be described as predominantly rural or urban. We are left with no solid basis for hypothesizing that rural and urban residents will differ from one another.

Despite the unclear expectations, it seemed reasonable to investigate whether there were differences in attitudes toward oil drilling or nuclear power based on place of residence. Two measures were used to test for these differences. First, I used the Census Bureau's code for metropolitan areas. The Census Bureau uses counties as the basic units in its metropolitan-area coding system. A county is a "metropolitan statistical area" if it has at least one city or "urbanized area" with a population greater than fifty thousand and if the total county population is greater than a hundred thousand.[58] Second, I created a measure in which all counties with populations greater than six hundred thousand were coded as urban, and counties with lesser populations as rural. This stricter standard probably comes closer to what most people would consider the key rural-urban distinction.

The results of this examination were a series of findings of "no difference." None of the measures of rural-urban differences was associated with any attitude differences. Although no differences appeared in simple tables, a rural-urban difference did appear when other factors were held constant in multivariate models (at the end of this chapter). As we shall see later, people living in counties with populations of six hundred thousand or more—that is, the major metropolitan areas of the state—were far more likely to oppose drilling for oil in government parks and forest reserves. Why they opposed oil drilling in parks, but not along the coast is not clear, but the result is no fluke. As we shall see later, it appears in both years for which we have data (1990 and 1998). Perhaps there is something about the label "parks" that is especially attractive to urbanites. Why no relationships exist with offshore oil drilling and nuclear power is not clear, but it seems reasonable to guess that unlike for many environmental issues, there is no special connection between coastal oil development or nuclear power and rural or urban areas. That is, these results fit quite well with Kowalewski's warning that the relationship between place of residence and environmental attitudes is far more complicated than the simple rural-urban distinction.

Energy, Party Identification, and Ideology

So far we have been focusing on demographic characteristics. These are not the only variables related to opinions on oil or nuclear power. Other variables—most importantly party identification, ideology, and values—are also related to environmental opinions. Because these other variables are also related to opinions on both economic and social issues, they do not help us distinguish between the two patterns, but they do tell us more about the sources of support for, and opposition to, increased energy development. In this section, we address party identification and self-identified ideology. We leave the more elaborate ideological explanations for the next section.

The central arguments about party identification and ideology are straightforward. The Republican Party represents business interests and wealthier segments of society; the Democratic Party represents the working class and lower-income segments. When there are benefits to be had from exploiting the environment, they disproportionately go to businesses and their wealthy owners and managers. When there are environmental costs to be paid—for example, in terms of polluted neighborhoods or unsafe working conditions—they tend to be paid by workers and the poor. On top of that, Republicans tend to be conservative and therefore to oppose government regulation of the business sector. Democrats, in contrast, tend to be liberal and therefore more open to government intervention in the marketplace. Because so many environmental controversies turn on government regulations and action, environmental issues trigger basic liberal and conservative views about the appropriate size of government and its role in society.

A few scholars argue that environmental issues, if properly understood, have nothing to do with partisan or ideological differences. Most prominently, Robert Paehlke argues that "environmentalism is inherently neither left nor right," that environmentalism generally "stands outside or beyond the left-right spectrum," and that ethical issues on which environmental policy debates turn are unrelated to the economic disputes that drive contemporary partisan and ideological disputes.[59] In the abstract, Paehlke's argument is certainly interesting, but it suffers from two serious weaknesses. First, Paehlke ignores social issues, treating conventional politics as if they focused only on economic questions. Paehlke's discussion of ideology and of the meanings of "Left" and "Right" refers only to economic questions, "class struggle," "union-management relations," "distributive" politics, and the like.[60] He completely ignores the many studies of both general public opinion and political leaders showing that a second dimension exists in American politics, a dimension dealing with issues that have a basis in morals and value judgments—issues such as civil rights, women's equality, free speech, abortion and birth control, pornography, lifestyles, sexual preferences, and related matters.[61] Having tossed out all noneconomic issues, Paehlke argues that environmentalism stands alone as the second dimension, as the top graph in figure 5.5 illustrates. Almost all public opinion researchers, in contrast, would say that the bottom graph in figure 5.5 captures the situation far more accurately. The question then becomes whether environmental issues

Figure 5.5   Hypothetical Dimensions of Public Opinion

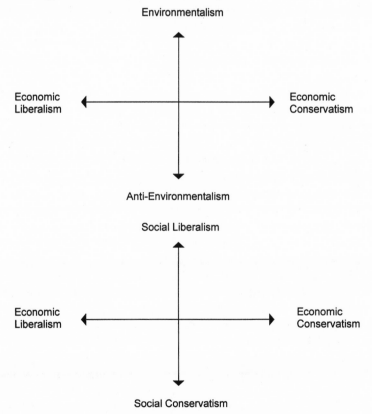

fit better with economic issues, with social issues, or as a third dimension. The evidence we have seen so far leans toward grouping environmental issues with social issues.

Second, Paehlke ignores studies of political leaders showing that Democratic and liberal leaders tend to take pro-environment stands, while Republican and conservative leaders tend to favor development.[62] Perhaps these partisan leaders ought not to think as they do. Perhaps this is a large-scale example of false consciousness, but the evidence is fairly clear. Environmentalism may have been a nonpartisan, nonideological issue when it first arose in the early 1960s, but by the late 1960s, when pollsters started asking questions about it, it had become both ideological and partisan among both the general public and political leaders. That is exactly what we find with our California data on energy issues.

As figure 5.6 shows, Republicans are more likely than Democrats to support offshore drilling. Although Republicans clearly hold views that are more pro-development than those of Democrats, the relationships are not perfect. Support for more energy development does not smoothly rise from the Democratic end to the Republican end of each party-identification scale. One reason might be that in some ways weak party identifiers are more committed to their parties than independents who say they only lean toward their parties, but in other ways the independent "leaners" seem more committed. That is, past research suggests there is some question about the sequence of the middle categories as one moves from the Democratic end of the scale to the Republican end of the scale. A second reason is that those in the middle are far less knowledgeable about politics than are strong identifiers on either end of the scale.[63] So when one compares strong party identifiers with independents, two things vary—party attachment and political knowledge. Nevertheless, despite the lack of a perfect fit, the evidence obviously indicates that Republicans support energy development far more strongly than Democrats do.

Figure 5.6 also shows that the relationships between opinions and party identification have weakened over time. In 1981, the differences between strong Democrats and strong Republicans ranged from 21 to 27 percent (depending on the energy source in question); by 1998, those differences had diminished to 17 to 23 percent. The decline is modest, but it fits the pattern found with age. As the RAS model predicts, because the news media have given less and less attention to energy issues, differences between pro-environment and pro-development groups have diminished.

We can gain further insights from the RAS model by exploring the relationships among opinions on energy issues, party identification, and political awareness. Partisanship is not as appropriate a predisposition as general environmentalism for purposes of using the RAS model, but it should work reasonably well. Partisanship has a huge influence on how people organize their

Figure 5.6   Support for Energy Sources by Party Identification

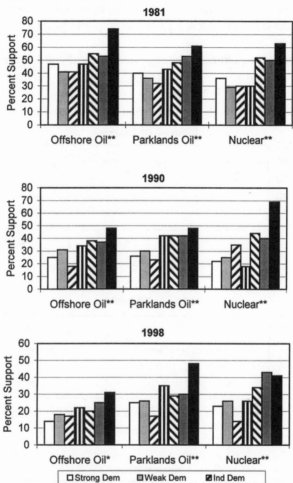

*p < .10; **p < .05

opinions about politics.[64] The news media often present news with partisan cues to help their audiences put issues in context. Political leaders, for example, are normally identified by party if they have ever held elective office. Because Democratic leaders are mostly pro-environment and Republican leaders are mostly pro-development, their statements about energy issues should help Democrats learn that they should be pro-environment and Republicans learn that they should favor development. To see whether this relationship behaves as the RAS model predicts, figure 5.7 separates Democrats, independents, and Republicans, and shows their views on oil drilling and nuclear power by their levels of political awareness.

In figure 5.7, the political awareness scale is actually the renamed political-

Figure 5.7   Energy Opinions by Party Identification and Awareness

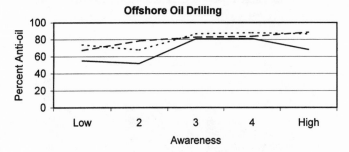

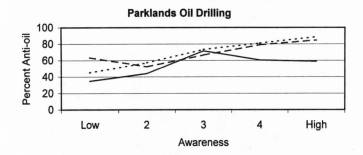

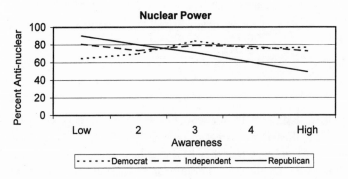

*p < .10; **p < .05

knowledge index used earlier in this chapter. Zaller maintains that the best measure of people's "cognitive engagement with politics"—how much they pay attention to and understand politics—is how much they know about politics. So, following Zaller, we measure political awareness with an index counting the number of correct answers people gave to five political-knowledge questions.[65]

The RAS model predicts that as awareness increases, Democrats should become increasingly anti–oil drilling and anti–nuclear power. Republicans, in contrast, should initially become more opposed to energy development as awareness increases, but then at upper levels of awareness should begin to recognize that their party generally supports energy development. That is exactly what we see. In all three panels of figure 5.7, large gaps appear between Democrats and Republicans among the most politically informed. The Republicans show the characteristic curve predicted by the RAS model. The RAS model thus yields another way of thinking about partisan differences on energy development. If it were not for those scoring highly (four or five) on the awareness scale, there would not be any significant differences between Democrats and Republicans in opinions about energy development. The partisan differences emerge from the best informed respondents.

We turn now to the influence of self-identified ideology. As one should expect, figure 5.8 shows that conservatives are more likely than self-identified liberals to support all three forms of energy production in all three of our samples. However, just as for age and party identification, the relationship weakens enormously over time. Take opinions about offshore oil drilling as an example. In 1981, 73 percent of conservatives favored more offshore drilling, but only 28 percent of liberals did—a difference of 45 percent. By 1990, the difference had fallen to 37 percent. By 1998, the difference was down to only 19 percent.[66] Over seventeen years, liberals and conservatives had come 26 percent closer together. This is the same pattern of weakening relationships that we saw with age and partisanship, but here it is far stronger.

The RAS model predicts that as awareness increases, liberals' opposition to energy development will rise and that conservatives' opposition will first rise, then fall. The data do not follow the predicted pattern as neatly as do party identification data, but the important elements of the pattern appear. In particular, at the upper levels of political awareness, liberals, moderates, and conservatives divide exactly as the model predicts. Liberals are the strongest environmentalists, while conservatives take the opposite stand, and moderates fall in between. As in the earlier examples, the upper end of the political knowledge/awareness scale is the primary source of the differences between the views of liberals and conservatives.

Figure 5.8 Support for Energy Sources by Ideology

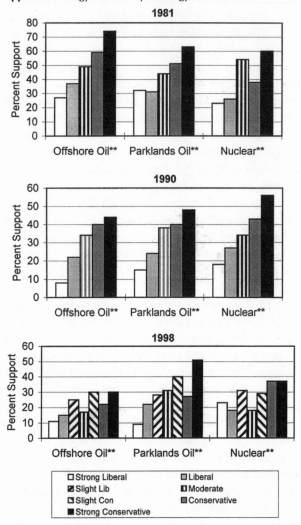

*p < .10; **p < .05

BEYOND IDEOLOGICAL LABELS: DOUGLAS
AND WILDAVSKY'S CULTURAL THEORY

There is far more to ideology than whether someone prefers the label "liberal," "moderate," or "conservative." Indeed, the answers to these questions reflect far less than a casual observer might expect. One of the most disturbing findings that has emerged from decades of research on public opinion is how little the public understands about the political debates raging among politicians and activists. Research on the subject has revealed that large numbers of people are remarkably poorly informed.[67] Of particular relevance here, repeated studies since 1940 have shown that only about half the public even understands what the terms "liberal" and "conservative" mean. As early as 1940, a Gallup poll found that only 45 percent of the public could explain the terms. In a series of studies conducted between the 1950s and the 1980s, that question was examined again and again, always with the same answer: Only about half the public understands reasonably well what the words "liberal" and "conservative" mean.[68]

The consequence of these findings for our study is that if we want to understand the influence of ideology on opinions toward energy development options, we need to look beyond ideological labels. We need to find other ways to measure people's basic values and beliefs.

An alternative to measuring people's ideological views by asking them what labels they prefer is to ask questions tapping people's feelings about the substantive issues that make up ideologies. For example, we can ask people whether they favor increasing economic equality (a central belief of liberalism) or reducing the size and influence of government (a core conservative goal). By creating indexes consisting of sets of substantive questions about ideological beliefs, we can measure the ideologies and values of people who do not use or recognize ideological labels. A number of scholars have pursued this approach. We will discuss two groups of investigators who have related their work to opinions on environmental issues.

In a series of books and articles beginning in the 1980s, Mary Douglas and Aaron Wildavsky developed "cultural theory" to explain why people accept or reject environmentalism and why they choose which potential hazards to fear and which to ignore.[69] Their theory, based in anthropological research, holds that patterns of social relationships are determined by two variables. The first, "group," is the extent to which people are incorporated into and controlled by communities or other social groupings. The greater the incorporation, the greater the group influence on individual decisions and the lesser the individual's choice. The second variable, "grid," is the extent to which external constraints limit individual choices and behaviors. The combinations of these two variables yield four patterns of social relationships and corresponding "worldviews," values that in various combinations characterize all societies.

Egalitarianism stems from high group control, but low external constraint. Egalitarians believe—as the label implies—in relative equality in the community. *Individualism* stems from low group control and few external constraints. Individualists believe that people should be on their own and not rely on others for material assistance. *Hierarchicalism* stems from high group control and strong external constraint. Hierarchicalists believe in strong social and moral guidance from their community leaders. *Fatalism* stems from low group control and strong external constraint. Fatalists see the world as threatening and uncontrollable but feel that they cannot turn to their community for help.

These four cultural biases or worldviews also yield four characteristic responses to hazards and threats in the world.[70] Individualists tend to see lower risks than others see, and individualists are far more likely than others to accept risks in exchange for economic returns. Cultural theory, therefore, predicts that these people would be most likely to support further energy development. Egalitarians are especially concerned with potential risks caused by what they see as inegalitarian institutions—big government and large corporations. They are also most likely to favor policies that reduce risks at the expense of economic growth. Moreover, egalitarians are especially concerned with potential risks caused by large corporations or big government because they stem from inegalitarian organizations. Consequently, these are the people who are most likely to fear nuclear power, offshore oil development, genetic engineering, and similar threats. Hierarchicalists fear threats associated with social or moral breakdowns—for example, war, terrorism, mugging, pornography, or AIDS. Fatalists basically fear everything, for to them the world is a mysterious and threatening place altogether.

Douglas and Wildavsky argue that every society has a mix of people with these four types of worldviews. When a particular worldview is especially prevalent in a society, we can describe the entire society in those terms, but in the United States there is a mix of at least three of the types (fatalism is rare).

A number of indexes have been developed to measure each of these worldviews, but I will examine only two in this study—individualism and egalitarianism. I will limit the discussion to these scales for two reasons. First, previous research has shown that egalitarians should be the most pro-environment (and consequently the most anti–energy development), and individualists should be the most pro-development. Second, our 1998 California data set includes a set of questions designed to measure these two worldviews, but not hierarchicalism or fatalism.

To construct the egalitarianism and individualism indexes, respondents were asked whether they agreed strongly, agreed somewhat, disagreed somewhat, or disagreed strongly with the following statements:

Egalitarianism Questions

1. The world would be a more peaceful place if its wealth were divided more equally among nations.

2. We need to dramatically reduce inequalities between the rich and the poor, whites and people of color, and men and women.
3. What our country needs is a fairness revolution to make the distribution of goods more equal.

Individualism Questions

4. Competitive markets are almost always the best way to supply people with the things they need.
5. Society would be better off if there were much less government regulation of business.
6. People who are successful in business have a right to enjoy their wealth as they see fit.

The indexes were built by assigning the numbers one to four to the four possible answers to each question, adding up the answers, and subtracting two so that the resulting indexes range from one to ten.[71]

The questions, borrowed from a study by Richard Ellis and Fred Thompson, state substantive issues in plain language.[72] Even someone who does not know what the word "liberal" means may have an opinion on whether our country "needs a fairness revolution to make the distribution of goods more equal." With questions such as these, we can take a second look at the influence of ideological views on opinions regarding energy development.

Because the egalitarianism and individualism measures are not in common use, I should note that the egalitarianism index has a fairly uniform distribution with a mean score of 6.0, skewed slightly to the pro-egalitarian side. The individualism index is skewed even more toward the individualist side, with a mean of 7.4 and only 7 percent of the respondents receiving scores of four or less. Clearly, both sets of ideas are popular, although the individualist views are substantially more popular among Californians.

Before looking at the data, we need to consider what we should expect to find. The answer about the direction of the relationship is plain: Individualists should favor energy development, and egalitarians should oppose it—as Douglas, Wildavsky, and others have argued. Indeed, it would be surprising to find anything else, given the content of the questions. These questions do not measure ideology by the conventional "liberal-conservative" or "Left-Right" labels, but they nevertheless tap recognizable aspects of liberal-conservative ideology. The questions on individualism could alternatively be described as questions about conservative, pro-business values—that is, they tap a central aspect of modern conservatism. The questions on egalitarianism could be described as questions about economic equality, a central aspect of modern liberalism. In other words, the cultural theorists took measures of ideology that are well known to public opinion researchers and provided different labels and theoretical explanations for them.[73] There is certainly nothing wrong with reinterpreting past research from a new theoretical perspective, but readers should keep in mind that "cultural theory" and old-fashioned liberalism and conservatism have much in common.[74]

Based on previous research, we can make another prediction about what we should find—namely, that the individualism and egalitarianism indexes should do a better job of explaining energy opinions than does self-identified ideology. This hypothesis comes from the work of Ellis and Thompson, who conducted surveys of both the general public and of activists in Sawmill County, Oregon, in an effort to test Douglas and Wildavsky's cultural theory. They found that the two indexes performed far better as predictors of environmentalism than did either ideology or party identification.

The surprise in figure 5.9 is that contrary to Ellis and Thompson, egalitarianism and individualism do *not* work better than ideology. To be sure, they are related to opinions on energy policy. Support for energy development generally increases with individualism and decreases with egalitarianism (the wide swings of support for energy development at low levels of individualism are not very important because of the small number of respondents). However, the relationships are about as strong as the relationships with ideology reported above.

More important, at least in this first look at the data, the individualism and egalitarianism scales do not seem to explain everything, as Wildavsky and Douglas claimed they would. Cultural theorists—especially Wildavsky—have argued that the rise of "radical egalitarianism" is a major cultural shift in the United States and explains the rise of the environmental movement.[75] The data in figure 5.9 suggest a tamer conclusion: Cultural worldviews help explain attitudes toward energy policy, but there is no sharp divide between egalitarians and individualists. Cultural theory does not seem to be the key that opens everything.

We explore the relationship further by turning once again to the RAS model. The data come out roughly as the RAS model predicts for the egalitarianism index in figure 5.10, but there are a couple of unexpected twists. In the case of attitudes toward offshore oil drilling, respondents scoring high on the index become increasingly anti–oil drilling (as the RAS model predicts) as political awareness increases. At the other end of the scale, people with low scores have the predicted ∩-shaped curve—opposition to oil drilling first increases and then decreases as awareness grows. The unexpected twist is that the anti-egalitarians peak at a higher level of support for oil than is found among the egalitarians. This pattern is repeated in attitudes toward oil drilling in parks and government forest reserves. Only in the case of nuclear power does the support of those on the low end of the egalitarianism index always fall below the level of those who score high on the index. Still, these are only minor deviations from the RAS

Figure 5.9 Support for Energy Sources by Individualism and Egalitarianism, 1998

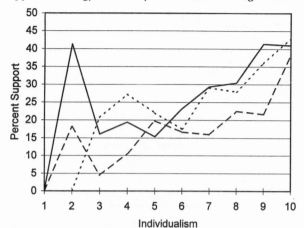

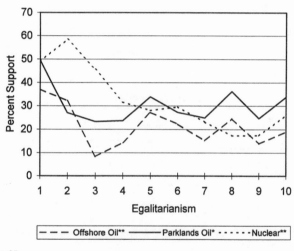

*p < .10; **p < .05

model's predictions. We will return to the question in the final section of this chapter.

In the case of the individualism index, the same basic pattern emerges for both types of oil drilling. As political awareness increases, people scoring on the low end of the individualist index generally become increasingly opposed to off-shore and parklands oil drilling. In contrast, as awareness increases, opposition

Figure 5.10   Energy Opinions by Egalitarianism and Awareness

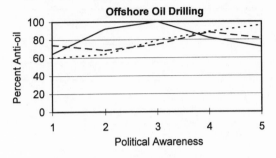

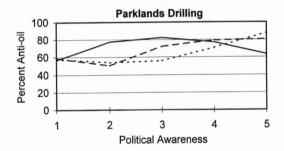

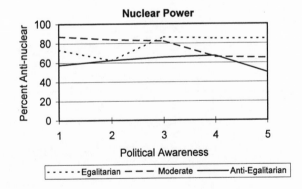

from individualists first increases and then declines in the predicted ∩-shaped relationship. Only in the case of nuclear power and individualism does the RAS model fail to give us much insight. Awareness seems to have little effect on anti-individualists, and the ∩-shaped curve fails to materialize for individualists. In sum, the RAS model helps explain the relationships we find here, but it fails, at least in these simple figures, to explain everything.[76]

Overall, the individualism and egalitarianism scales certainly help explain attitudes toward oil development and nuclear power. Yet just as certainly, they are not the complete explanation that Wildavsky, Douglas, and other cultural theorists have claimed them to be. With the addition of the political awareness/ knowledge indexes, however, their results become stronger. In the final section in this chapter, we will investigate them further and show how they perform when other causes are controlled. Before doing that, however, I turn to a second ideological approach that claims to explain people's attitudes toward environmental issues.

BEYOND IDEOLOGICAL LABELS: INGLEHART'S POSTMATERIALISM THEORY

A second theoretical approach to measuring people's ideologies and values without relying on simple labels has been developed by Ronald Inglehart.[77] Drawing on the socialization literature and the work of psychologist Abraham Maslow, Inglehart argues that people develop values and assign priorities to them during their formative years, from childhood into early adulthood.[78] Moreover, because of the changing nature of the times across the twentieth century, people from different generations develop different values. People who grew up during the early part of the century were socialized during times of widespread poverty (the Depression) and war (World Wars I and II). People who grew up in the postwar generation, however, were socialized during a period of unprecedented prosperity. They grew up without having to worry about basic material needs, such as whether they would have enough to eat or would be required to go off to war. As a result, Inglehart claims, prewar generations tend to emphasize material values, while postwar generations emphasize postmaterial values—freedom, self-expression, and quality of life.

Studies of value priorities measure them by asking people to rank-order four goals in terms of importance—fighting rising prices, maintaining order in the nation, giving people more say in government decisions, and protecting freedom of speech. The first two goals are considered *materialist;* the second two are *post-materialist.* The rankings are combined to construct a four-point scale. People who rank the two postmaterial goals as most important are given the score of four; people who rank a postmaterial goal as most important and a material goal as second most important are scored as three; those who rank a material goal as most important and a postmaterial goal second are given a two; and those who rank both material goals as most important are scored as one.

In a few studies, a more elaborate list of twelve questions is used, but most studies (including this one) rely on the four items above.[79] In the California survey, the population is skewed slightly to the postmaterial side, with an average score of 2.8 on the one-to-four scale.

Although the value questions do not specifically ask about environmental issues, Inglehart and others maintain that environmentalism is a postmaterial value. After attaining sufficient material benefits, people develop a taste for a cleaner, safer, more beautiful environment. In fact, several studies of value priorities have focused on attitudes toward nuclear power. The implication is quite clear for our purposes: Materialists should tend to support oil development and nuclear power, while postmaterialists should tend to oppose them.

The data, unfortunately, offer little support for Inglehart's claims insofar as energy policy is concerned. Support for oil drilling and nuclear power varies only a few percentage points between materialists and postmaterialists. None of the relationships is statistically significant, That is, the differences could easily be the result of random sampling error. Breaking these relationships down by political awareness to see if differences appear among the most informed does nothing to improve the situation. The data reveal that if anything, knowledgeable materialists are slightly more likely than postmaterialists to *oppose* offshore oil drilling and nuclear power, although these differences are not quite statistically significant. This is the opposite of what Inglehart's theory predicts. In the case of drilling for oil in parks and forest reserves, well-informed materialists and postmaterialists share virtually identical views.

The findings so far indicate that Inglehart's theory of the generational change from materialism to postmaterialism offers no insight into why people support or oppose energy development. We will return to the possible influence of value priorities in the final multivariate assessment of the causes of opinions, but our first look has clearly not been successful.

Before proceeding to a final assessment of the causes of people's opinions on energy options, it will be useful to offer a brief overview of the discussion so far. The correlations shown in table 5.3 summarize the data presented in this chapter. Two useful observations can be made. First, as just noted, the correlations between individualism and the three energy questions are among the strongest in the table, but they are not any larger than the correlations with either party identification or self-identified ideology. The correlation between egalitarianism and attitude toward nuclear power also ranks among the strongest, but egalitarianism works less well with attitudes toward the two types of oil development. In short, the data show that the "cultural" variables of Wildavsky and Douglas perform well, but do not stand out above party identification or ideology—or, in some years, age and gender.

Second, none of these correlations is very strong. They all range from moderate to weak in strength. This pattern is consistent with the findings of other studies that have examined attitudes toward environmental issues.[80] It is also

Table 5.3　Pearson Correlations Coefficients between Energy Opinions and Sources of Opinions

	Offshore Oil			Parklands Oil			Nuclear Power		
	1981	1982	1983	1981	1990	1998	1981	1990	1998
Income	−.02	−.02	−.05	−.02	−.03	−.05	.13**	.11**	.13**
Education	−.10**	−.14**	−.09**	−.07**	−.15**	−.10**	.04**	.13**	.08**
Information			−.10**			−.17**			.13**
Age	.20**	.14**	.04	.15**	.16**	.02	.21**	.21**	.07**
White	.03	−.06	−.05	−.01	−.10**	−.12**	.17**	.11**	.10**
Black	.01	.06	−.04	.00	.06	.02	−.11**	−.11**	−.04
Asian	.00	.02	.00	−.02	.04	.02	−.05	−.04	−.03
Latino	.02	.03	.01	−.04	.05	.11**	.08**	−.05	−.07
Gender	−.06*	−.03	−.12**	−.03	−.03	−.11**	−.17**	−.13**	−.20**
Oil County	−.03	.01	.05	.03	.01	−.03			
Nuclear County						.02			.05
Urban			−.01			−.05			.00
Party	.19**	.16**	.16**	.16**	.18**	.14**	.25**	.29**	.21**
Ideology	.31**	.19**	.16**	.21**	.17**	.21**	.27**	.22**	.13**
Individualism			.21**			.19**			.20**
Egalitarianism			−.11**			−.05			−.25**
Materialism/Postmaterialism			−.06			−.06			−.01

Source: Data from Field Polls 8104 and 9004, and from the California Offshore Oil Drilling and Energy Policy Survey.
*.05 < p < .10; **p < .05
Note: Sample sizes for correlations: 1981 data: 549–585; 1990 data: 915–968; 1998 data: 526–764.

typical of studies of social issues. In general, opinions on economic issues are more closely aligned with party identification, ideology, and the demographic variables we have been discussing. Because social issues cut across traditional party and ideological cleavages in our society, their relationships are not as strong. As we shall see in the next section, this holds even when examining the combined effects of the explanatory variables.

MODELING ATTITUDES

The weakness in our discussion of the sources of opinions on energy issues so far is that we have been addressing one variable at a time. First we looked at the effects of income, then education, and so on. Do they collectively add up to a good explanation, or do the pieces overlap so that we were explaining the same variance over and over? The only way to find out is to put all the pieces of the explanation together into a single multivariate model and see how well they work.

To do this, we need to estimate a set of ordinary least-squares regression models. These models allow us to isolate the effects of each of the variables discussed earlier in this chapter, while controlling for all the effects of all the others. In addition, these models allow us to explore how Zaller's RAS model and Douglas and Wildavsky's cultural theory work together to explain attitudes toward energy policy. We have seen a few figures describing some of the results predicted by the RAS model, but the RAS model predicts interactions among variables that cannot easily be seen without examining the effects of several variables simultaneously. In short, multivariate models allow us to develop a far better understanding of what causes environmental opinions than do single-variable methods.

One might describe these results in several ways. I will approach the task in two stages. I will begin by examining a set of regression equations over time from 1981 to 1998, so that we can see how the causal relationships have changed. Then I will examine the 1998 data in more detail, exploring how the RAS and cultural theories combine to explain attitudes toward energy production.

Before discussing the data, I must briefly discuss the dependent variables and say a few words about how the independent variables are scored. In the figures earlier in the chapter, attitudes were described as simply pro- or anti-oil or nuclear power. In fact, the questions actually had a little more detail, and I use it here. In 1990, respondents were asked to agree or disagree with the three statements about energy questions that were quoted in the first section of this chapter. By recoding those who said they did not know what to think at a midpoint between pro- and anti-energy, we produce a three-point scale. In 1981 and 1998, the questions asked respondents whether they agreed strongly, agreed somewhat, disagreed somewhat, or disagreed strongly. Again, by coding in the middle those who said they were not sure, we have five-point scales for those two years.[81]

The independent variables are coded as described earlier in this chapter, with the exception of income. The Field Institute changed its income categories over the years. In the 1981 survey, the question used zero to ten thousand dollars as its lowest income category. In 1990, zero to twenty thousand dollars was the lowest category, and the higher categories were multiples of twenty thousand dollars, with the top category being sixty thousand or more. In 1998, the question changed again, with the addition of a sixty-to-eighty-thousand-dollar category and a new top category of eighty thousand or more. In addition, because of inflation, the interpretation of income has changed over time. These differences are not important because income has so little effect (except in the case of nuclear power), but readers should be careful when comparing the effects of income over time. The rest of the coding details are in the data appendix.

With these preliminaries out of the way, we can turn to the equations in the models explaining attitudes toward offshore oil drilling in table 5.4. As our initial look at income suggested, income has no significant effect on attitudes toward offshore oil development. In 1981, the coefficient of 0.03 indicates that

Table 5.4   Regression Models of Attitudes toward Offshore Oil Drilling

	1981 Coefficient	1990 Coefficient	1998 Coefficient
Income	0.03*	−0.02	−0.03
Education	**−0.13****	**−0.06***	−0.07
Age	**0.18****	**0.09****	0.05
Black	**0.37***	**0.38***	0.01
Asian	0.21	0.26	0.31
Latino	−0.25	**0.23***	**0.28****
Female	**−0.20****	−0.07	**−.20****
Oil County	−0.23	0.11	0.11
Urban		0.00	−0.07
Party ID	**0.08****	**0.06****	**0.08****
Ideology	**0.31****	**0.09****	**0.07****
Intercept	**1.62****	**1.13****	**1.68****
N	906	524	749
Adjusted R²	0.13	0.06	0.05

Note: **Bold** coefficients are at least of borderline statistical significance (p < .10).
*p < .10; **p < .05

as income increased by one category (for example, from $0–10,000 to $10–20,000), respondents scored an average .03 higher on the five-point, pro-drilling scale. To put it another way, the difference between the richest and poorest Californians was only 0.12—a substantively and statistically insignificant difference. The results were essentially the same in 1990 and 1998. Family income has no effect on attitudes toward oil drilling along the California coast.

Education does have an effect on attitudes. As expected, the better educated one is, the less likely one is to support offshore oil drilling. The effect of education weakens over time, however. In 1981 it is statistically significant, but in 1990 the effect is weaker and of only borderline significance ($p < .08$). By 1998, the effect is still weaker, and this time it is not significantly different from zero. When the information scale is added in the expanded 1998 equation, which I discuss later, the effect of education shrinks even further.

Age also has the expected effect in this model—the old are more likely to support offshore drilling than the young—and as with education, its effect diminishes over time. In 1981, age had one of the strongest effects of any variable. This can be seen in two ways. The coefficient of 0.18 indicates that for every ten years of age, the typical respondent is 0.18 units more likely to support oil drilling on the five-point scale. That is, the difference between a twenty-year-old and a seventy-year-old is almost a full unit change on the attitude scale (for example, from undecided to moderately pro-drilling). Age has the second-largest impact of any variable in 1981. By 1981, however, the influence of age had weakened to the point that there were only small and insignificant differences between the young and old. It was no longer one of the dominant effects in the model.

The next three variables in the model distinguish blacks, Asians, and Latinos from whites. That is, these coefficients should be interpreted as showing the differences between a group and the excluded, or baseline category, of whites.[82] In 1981, for example, the black coefficient of 0.37 indicates that blacks scored 0.37 higher than whites on the drilling support scale. In fact, in both 1981 and 1990, blacks are more likely than whites to support additional offshore oil drilling (although the 1981 difference is only of borderline significance at $p < .06$). By 1998 the difference vanishes. For their part, Latinos lean against coastal oil drilling in 1981, but they lean in favor of it in 1990 and 1998. Moreover, the 1990 and 1998 results are large enough to be statistically significant. Asians are consistently a bit more in favor of more oil drilling than are whites, but the coefficients never reach statistical significance. In sum, race and ethnicity contribute to our understanding of energy attitudes, but the patterns change over time.

Women are less likely than men to support offshore oil development in both

1981 and 1998, but not, curiously, in 1990. This breaks the general pattern of relationships fading away over time.

Proximity—that is, living in Santa Barbara or Ventura Counties, counties with both offshore oil production and vociferous, anti-oil political movements—made no difference. In 1981, the coefficient is negative, which indicates a slight but statistically insignificant inclination against oil. In both 1990 and 1998, however, the coefficients are positive, indicating a similarly slight inclination in favor of oil. On balance, the only reasonable conclusion that one can draw from these data is that living in the general area of offshore oil production has no effect on one's attitudes toward it. If a NIMBY effect does exist, it presumably exists on a much smaller scale. One might possibly find NIMBY reactions to proposed oil facilities within, say, a mile or two of one's home, for example. Whether neighborhood NIMBY reactions occur or not, the evidence here indicates that they are not a widespread phenomenon, at least insofar as offshore oil drilling is concerned.

Living in a metropolitan county made no difference to people's attitudes toward offshore oil drilling. Although some studies have found urbanites to be more pro-environment, that pattern failed to appear in these data.

The effects of party identification are quite consistent over time. It has a substantial influence on opinions, and the size of the influence does not appear to change from year to year. This conclusion differs from the one drawn from the simple bivariate data shown in figure 5.6. The data in that figure indicate that the difference of party identification declined over time. Here we see that when we control for the effects of other variables in the regression model, the effects of party identification did not change.

Unlike party identification, the effects of self-identified ideology do change over time. They get smaller. In 1981, ideology had more influence than any other variable, but by 1998, its influence had been sharply reduced. Ideology still had an important and statistically significant effect in 1998, but the conclusion that attitudes toward oil development were less ideological than they had been during the last oil crisis is hard to dispute.

The final useful observation to make about these models of attitudes toward offshore oil drilling is that the adjusted R-square, the measure of how well the equations explain the dependent variables, declines over time. In 1981, the equation explains 13 percent of the variance; by 1998, the equation explains only 5 percent of the variance. We need to be careful interpreting the decline in the explanatory power of these models over time. Several factors—including changes in the distribution of the dependent variable—can cause this number to fall.[83] In this case, however, it seems that the decline reflects the social process described by the RAS model: people forgetting the connection between their pro-environment or pro-development values and the issue of offshore oil development. As oil issues faded from the nation's front pages, public differences of opinion faded as well.

The results of the models explaining attitudes toward oil drilling in government parks and forest reserves, presented in table 5.5, are generally similar to those for attitudes toward offshore oil drilling. Income has no effect. The more years of education one had in 1990, the less likely one was to support oil drilling, but in 1981 and 1998 the coefficient did not reach statistical significance. The results for age match those for offshore drilling. Age had a substantial effect in 1981, but the effect so diminished over time that by 1998 the old were no longer significantly more likely than the young to support drilling.

The black, Asian, and Latino coefficients differ from the offshore oil pattern. Here we see all three coefficients becoming more positive. Because they are measured in comparison with the baseline of whites, this means that all three groups were becoming more pro-oil drilling than whites. But given that the entire population is moving away from support for oil drilling, it would be more accurate to say that whites are moving in an anti-oil direction more quickly than the other groups.

The coefficients for women and proximity show little change over time. Women are only a slight, insignificant bit more anti-oil than men, and the difference does not change over time. Similarly, living in an oil-producing county makes no difference to one's opinion in any year. Living in a large urban area,

Table 5.5   Regression Models of Attitudes toward Parklands Oil Drilling

	1981 Coefficient	1990 Coefficient	1998 Coefficient
Income	0.01	−0.01	−0.02
Education	−0.07	**−0.10****	−0.06
Age	**0.14****	**0.10****	0.02
Black	0.31	**0.41****	**0.44***
Asian	−0.01	**0.42****	**0.47****
Latino	**−0.52****	**0.28****	**0.53****
Female	−0.13	−0.05	**−0.18***
Oil County	0.21	0.01	0.00
Urban		**−0.19***	**−0.28****
Party Id	**0.10****	**0.07****	**0.07****
Ideology	**0.18****	0.05	**0.14****
Intercept	**1.91****	**1.35****	**1.70****
N	926	530	749
Adjusted R^2	0.07	0.09	0.08

Note: **Bold** coefficients are at least of borderline statistical significance (p < .10).
*p < .10; **p < .05

however, does influence attitudes. City dwellers were less likely to favor drilling for oil in parks and forests. Party identification consistently had a large, significant effect, and ideology also influenced opinions in both 1981 and 1998. In sum, there are a few deviations from the results we found for attitudes toward offshore oil drilling, but there are not many, and they are not important.

Table 5.6 presents the results for our model of attitudes toward nuclear power. Here we find a few differences worth noting, beginning with income. Unlike the results with oil, income has a positive, significant effect on support for nuclear power in 1981. Education only rises to borderline statistical significance in 1990, but its effect is the opposite of its effect on attitudes toward oil drilling. Instead of higher education pushing people toward the pro-environment side, here we see it pushing people toward the pro-development side. The effects of age are similar to those for our other dependent variables. Age is one of the strongest variables in 1981, but by 1998, its influence diminished enormously.

Race and ethnicity also diminish in their influence. In 1981, blacks and Asians leaned against nuclear power (although not at a significant level), but in 1998 the coefficients indicate that race and ethnicity have nothing to do with attitudes on nuclear power. These results, incidentally, do not match those of the simple bivariate data, which show that blacks, Asians, and Latinos were roughly 10 percent less likely to support nuclear power and that those differences were of at least borderline significance. Here we see that those differences

Table 5.6   Regression Models of Attitudes toward Nuclear Power

	1981 *Coefficient*	*1990* *Coefficient*	*1998* *Coefficient*
Income	**0.13****	0.05	0.05
Education	−0.03	**0.07***	0.05
Age	**0.20****	**0.12****	**0.03**
Black	−0.33	−0.12	0.05
Asian	−0.46	−0.21	−0.27
Latino	−0.09	0.06	−0.04
Female	**−0.54****	**−0.21****	**−0.54****
Nuclear County		0.05	**0.41****
Urban		0.03	0.07
Party ID	**0.12****	**0.10****	**0.08****
Ideology	**0.21****	0.07	0.04
Intercept	**0.82****	**0.50****	**1.63****
N	920	529	749
Adjusted R²	0.17	0.14	0.07

Note: **Bold** coefficients are at least of borderline statistical significance (p < .10).
*p < .10; **p < .05

were the products of other variables included in the model—income, education, and so forth. Once we control for these variables, the racial and ethnic differences disappear.

The results for women match the bivariate data quite well. In all three years, women were consistently more opposed than men to nuclear power. Moreover, these are large differences. There is clearly something about nuclear power that women dislike. As I suggested above, the most likely explanation seems to be the threat of harm to people, which oil drilling does not pose.

Living in a county that has a nuclear power plant does make a difference in attitudes, but not in the direction that many might expect. People who live in the vicinity of nuclear power plants think there ought to be *more* of them. Although one might expect NIMBY responses to nuclear power, we find the opposite.

Party identification consistently has a large effect on attitudes toward nuclear power, with Democrats being most likely to oppose it, but ideology fades over time. In 1981, liberals were significantly more likely than conservatives to oppose nuclear power, but in 1990 and 1998 the coefficient failed to attain statistical significance. The issue of nuclear power faded from the nation's news media, and differences of opinion faded along with the news coverage.

To sum up, what we see in these equations is that most of the variables that worked in the bivariate tables also work here. Age, education, gender, party identification, and ideology generally influence attitudes on energy policy. Race and ethnicity have some effect. In particular, whites seem to be moving in a pro-environmental direction faster than other groups. Proximity to nuclear power plants seems to cause people to like them. Living in an urban area leads people to dislike oil drilling in parks. Income generally does not have much influence. Last, and perhaps most important, the models for offshore oil drilling and nuclear power—the two controversial types of energy production—lose more than half of their explanatory power (adjusted R-square) from 1981 to 1998. Only the model for attitudes toward oil drilling in parklands and forest reserves remains unchanged. Yet this question has never been controversial and has never been a routine subject of front-page newspaper stories or television news, as offshore oil drilling and nuclear power have. In other words, the diminishing predictive power of the statistical models corresponds to the diminishing attention the issues receive in the news media.

The models in tables 5.4 through 5.6 show us how attitudes changed over time, but they tell us nothing about the roles of political knowledge or cultural worldviews. Only the 1998 data set includes the variables that will allow us to investigate those relationships. We turn to that task now.

THE RAS MODEL AND CULTURAL THEORY

A brief recap of the RAS model will help set the stage for our second group of regression models. The basic propositions of the RAS model are: (1) exposure to mass-communication messages increases with political knowledge; (2) among those who have been exposed, acceptance of the message increases with knowledge if the message agrees with one's predispositions; and (3) among those who have been exposed, acceptance of the message decreases with knowledge if the message is contrary to one's predispositions. So acceptance of a persuasive, controversial message depends on the individual's political knowledge and on whether the message agrees or disagrees with his or her basic values. The predispositions we use here are the two cultural worldviews from Douglas and Wildavsky's theory—egalitarianism and individualism.

The core of our test of the combined RAS and cultural theories lies in the effects of the cultural value indexes and knowledge. The essential elements of the model should be captured by the following equation:

$$\text{Opinion} = b_0 + b_1(\text{Egalitarianism}) + b_2(\text{Individualism}) + b_3(\text{Knowledge}) \\ + b_4(\text{Egalitarianism} \times \text{Knowledge}) \\ + b_5(\text{Individualism} \times \text{Knowledge})$$

That is, cultural theory says that egalitarianism and individualism should both influence attitudes toward energy development. Egalitarians should consider it risky and oppose it; individualists should find it safe and support it. The RAS model says that as knowledge increases, the effects of egalitarianism and individualism should also increase. Those with little political knowledge should not recognize the connection between their values and their opinions on offshore oil drilling and nuclear power, but those with considerable political knowledge should recognize the connections and should bring their opinions in line with their worldviews. This means that there are interactions between the worldview and knowledge indexes. In the equation above, the interactions are assumed to be multiplicative. That is, the last two variables were created by multiplying the values of egalitarianism and knowledge, and individualism and knowledge.

Unfortunately the model specified above suffers the weakness that including more than one knowledge-interaction term increases the likelihood that "multicollinearity" among the variables will hide any effects. Multicollinearity is a statistical problem that occurs when two or more variables in a regression equation are so strongly correlated with one another that disentangling their effects becomes impossible.[84]

The situation with individualism is particularly poor. The individualism-knowledge interaction term is strongly correlated with knowledge ($r = 0.81$)

and individualism (r = 0.59). Such high correlations cause large standard errors in regression models so that one cannot tell whether or not a particular variable has an effect. The problem is compounded by the low reliability of the individualism index. The individualism index has a good deal of random measurement error.[85] As a consequence, when both the individualism-knowledge and egalitarianism-knowledge terms are included in any model, none of the terms achieves statistical significance. To test the model, therefore, we must choose which of the two interaction terms to include in each model.

Whether because of multicollinearity or because of a poorly specified model, the individualism-knowledge interaction term fails to work in the oil-drilling and nuclear-power models. Curiously, the situation was reversed in the equation explaining attitudes toward drilling in parklands and government forest reserves. The individualism-knowledge term worked, but the egalitarianism-knowledge term had no effect.

Multicollinearity and the low reliability of the individualism index may be causes of the failure of the individualism terms, but another possibility is that something about the issues themselves causes this pattern of findings. Both offshore oil drilling and nuclear power became highly publicized political battles, while the question of whether to drill for more oil in government parklands and forest reserves never gathered much attention. This suggests that egalitarianism is the relevant predisposition for environmental struggles (as Wildavsky claims) and that individualism is triggered when the dispute is about government regulation, with no obvious environmental values at issue. This line of reasoning seems persuasive, but unfortunately the limited data available do not allow us to sort out which answer is correct. In the tables below, we see the best model for each of the dependent variables. Further research will be needed to explain the roles of individualism and egalitarianism with more certainty.

In addition to the RAS/cultural theory variables, the regression equations include the variables examined in tables 5.4 through 5.6. If we are going to sort out the effects of knowledge and cultural values on opinions, we need to be sure that we are not accidentally looking at the effects of age, education, gender, or some other variable. By including these variables in the model, we can examine the effects of the cultural and RAS models while holding all other effects constant.

In tables 5.7 through 5.9, we present the results of four regression equations for each of the dependent variables. The first equation includes only the demographic variables. The second adds party identification, ideology, and political knowledge. The third adds the individualism and egalitarianism indexes. The final equation adds the egalitarianism-knowledge interaction term.

The point of presenting the four versions of each model is to show the relative contributions of each set of variables as the model becomes increasingly comprehensive. In the models above, we sought to show how relationships changed over time. Here we focus on showing the relative strengths of the different explanatory variables. This approach should yield a clear picture of how well the RAS and cultural models work and how much better they perform than the collection of demographic variables.

The first demographic equation for offshore oil drilling, shown in table 5.7, yields typical findings for a model explaining attitudes toward environmental issues. Older respondents are more likely than younger respondents to support additional drilling (as indicated by the positive coefficients). The well educated are more likely to oppose it than the poorly educated, and women lean against it more than men. Income, race, ethnicity, living in an oil-producing coastal county, and living in an urban area fail to achieve statistical significance. Overall the model performs poorly, explaining only 2 percent of the variance.

In the second equation, we see that adding party identification, ideology, and political knowledge slightly reduces the influence of the demographic variables. Age is reduced to borderline significance (p < .10); education becomes insignificant; and gender weakens but remains statistically significant. Party identifi-

Table 5.7   Regression Models of Attitudes toward Offshore Oil Drilling, 1998

Variable	(1) b	(2) b	(3) b	(4) b
Intercept	2.13***	1.80***	1.52***	.06
Income	−.01	−.02	−.05	−.04
Education	−.09**	−.06	−.04	−.05
Age (decades)	.06**	.05*	.01	.01
Black	−.12	−.03	.18	.14
Asian	.21	.26	.45**	.42**
Latino	.20	.23*	.24	.17
Women	−.27**	−.21**	−.15	−.15
Oil county	.10	.10	.00	−.03
Urban	−.07	−.07	−.06	−.01
Party ID		.08**	.08**	.06**
Ideology		.06**	.04	.03
		(.03)	(.04)	(.04)
Knowledge		−.05	−.06	.36***
Individualism			.12***	.10***
Egalitarianism			−.01	.24***
Material/postmaterial			−.11*	−.11**
Egalitarianism × Knowledge				−.06***
Adjusted R²	.02	.05	.10	.12
N	756	749	573	573

*.05 < p < .10; **p < .05; ***p < .01

cation and ideology both have significant impacts on opinions on oil drilling. The knowledge index fails to achieve significance, exactly as Wildavsky and others would have predicted.

One might suspect that including both education and knowledge would create a multicollinearity problem, causing the two variables to fail to achieve significance. However, even when each variable is included separately, without the other, it fails to achieve the usual $p < .05$ significance level. Both fall in the less persuasive $p < .10$ range. This suggests that there is some multicollinearity, but not a great deal. The more important result here is that with party and ideology in the equation, education and knowledge have only marginal effects. Moreover, the contribution of party identification and ideology increases the explained variance to 5 percent.

In the third equation, adding the individualism, egalitarianism, and postmaterialism indexes pushes age, education, and gender into statistical insignificance. The effects of party identification and ideology remain unchanged. Of the two cultural bias indexes, only individualism is statistically significant—and as cultural theorists would predict, it is positive. Individualists support oil development. The postmaterialism index also achieves borderline statistical significance (p. $< .10$), and the sign is in the correct direction. Moving up the scale of postmaterialism causes support for oil drilling to decline. Because of the contribution of the three new value indexes, the adjusted R-square doubles, to 10 percent.

In the fourth equation, the adding the egalitarianism-knowledge interaction term causes a good deal of change. The demographic variables remain small and statistically insignificant (with the exception of Asian, which is of borderline significance). Party identification diminishes slightly, but remains significant. Individualism, egalitarianism, knowledge, and their interactions all have large and significant effects. What may seem surprising is that the egalitarianism has a positive effect. Yet the egalitarianism-knowledge interaction term has an offsetting negative coefficient. The egalitarianism coefficient of 0.24 may seem larger than the interaction coefficient of -0.06, but the impact of the interaction is larger. Recall that the egalitarianism index has a one-to-ten scale, and the knowledge scale has a one-to-five scale. The interaction term, therefore, has a $5 \times 10 = 50$ point scale. So the potential impact of the interaction term is greater than the potential impact of the egalitarian scale alone. In other words, the more egalitarian a person is, the more he or she opposes offshore oil drilling.

Because visualizing the collective impact of egalitarianism, knowledge, and their interaction is difficult, figure 5.11 presents a three-dimensional histogram illustrating the effects. The axes on the "floor" of the figure are the egalitarianism and knowledge scores. The height of the histogram indicates level of support

Figure 5.11   The Joint Effects of Knowledge and Egalitarianism on Offshore-Oil Opinions

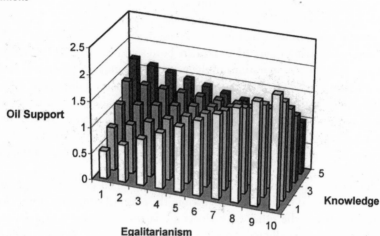

for offshore oil drilling. The floor is arbitrarily set at zero to make the figure easier to read.[86]

In the low-egalitarianism category (on the left side of the figure), as knowledge increases, support for offshore oil drilling increases as well. In contrast, among those with the highest egalitarianism scores (on the right side of the figure), as knowledge increases, support for offshore oil drilling declines. The cumulative result is that those with either high knowledge and low egalitarianism *or* low knowledge and high egalitarianism are the strongest supporters of offshore oil drilling. Those in the other two corners are the strongest opponents of offshore oil. This is the set of responses predicted by the RAS model.

Three final observations can be made about the offshore oil development equations. First, knowledge increases support for offshore drilling—once the effects of egalitarianism and individualism are taken into account. This is what scientists want: political awareness and knowledge leading people toward the scientists' views on safety issues. Yet it is not something we see unless we control for cultural predispositions and include the appropriate interaction terms. Second, the coefficients for individualism, egalitarianism, knowledge, and their interaction do most of the explaining in this model. Only party identification adds any independent explanatory power, and its coefficient is relatively modest. The other variables—including the postmaterialism index—contribute either not at all or only modestly to the model's ability to explain attitudes. In light of the fact that egalitarianism, individualism, and general political knowledge are not obviously connected to attitudes toward offshore oil drilling—or for that matter, to any environmental issue—this is a strong finding.

Table 5.8 reports the same set of equations for attitudes toward drilling for oil in government parklands and forest reserves. In the first (demographic) equation, gender and living in an urban area are both significant causes. As we shall see in the nuclear-power questions in the next table, parklands oil drilling is the only question for which living in urban or rural areas makes a difference. Urban residents like their parks without oil rigs, although there are no rural-urban differences for either coastal oil drilling or nuclear power. Education reaches only borderline significance. As with the offshore oil drilling model, this demographic equation does a poor job of explaining attitudes, as reflected in the adjusted R-square of 0.03.

In the second equation, relatively little changes. When party, ideology, and knowledge are added in the second equation, the effect of education drops close to zero. Otherwise, the other coefficients are largely unchanged. Republicans and conservatives generally tend to support oil, as the positive coefficients show. The well informed tend to oppose oil.

When the three value indexes are added in the third equation, we see the

Table 5.8   OLS Regression Models of Attitudes toward Parklands Drilling, 1998

Variable	(1) b	(2) b	(3) b	(4) b
Intercept	2.55***	2.26***	1.84***	3.20**
Income	.01	.01	−.05	−.05
Education	−.10*	−.01	.01	.01
Age (decades)	.05	.04	.02	.02
Black	.31	.31	.67**	.64**
Asian	.33	.32	.43*	.44*
Latino	.44	.38	.44**	.46**
Women	−.28***	−.22**	−.12	−.13
Oil county	−.02	−.03	−.13	−.13
Urban	−.29**	−.27*	−.34**	−.34**
Party ID		.07**	.06	.05
Ideology		.13***	.13***	.12***
Knowledge		−.17***	−.13***	−.50***
Individualism			.09**	−.10
Egalitarian			−.01	.00
Material/postmaterial			−.02	−.02
Individualism × Knowledge				.05**
Adjusted R²	.03	.09	.11	.12
N	756	749	573	573

*.05 < p < .10; **p < .05; ***p < .01

coefficients for being black, Asian, or Latino increase into statistical significance. All three groups favor more oil development in parks and government forests. This indicates that there are clear racial and ethnic differences in the distributions of the value indexes—especially individualism, which is the only significant effect.

The final equation in this set adds the individualism-knowledge term to the equation. The knowledge coefficient takes a huge jump in magnitude. For every one-point increase on the knowledge scale, respondents move a half point in the anti-oil direction. Individualism now shows a negative coefficient, but the individualism-knowledge interaction is positive. Just as with the offshore-oil-drilling model, because the interaction term has a range of one to fifty, this is a huge effect. Overall, the results match those of the offshore-oil-drilling equations, except that this time knowledge, individualism, and their interaction are the driving forces behind people's opinions.

The equations for nuclear power appear in table 5.9. In the first, demographic equation, only gender and living in a county with a nuclear power reactor are statistically significant. Women are much more likely to oppose nuclear power than men, and—contrary to popular beliefs about NIMBY effects—people living near nuclear power plants like them more than people living elsewhere in

Table 5.9   OLS Regression Models of Attitudes toward Nuclear Power, 1998

Variable	(1) b	(2) b	(3) b	(4) b
Intercept	2.01***	1.65***	1.78***	.84
Income	.08*	.05	.00	.00
Education	.03	.05	.09	.09
Age (decades)	.04	.03	−.01	−.01
Black	−.07	.05	.07	.05
Asian	−.35	−.27	−.38	−.40
Latino	−.10	−.04	−.11	−.16
Women	−.60***	−.55***	−.48***	−.48***
Nuclear county	.45***	.42***	.44***	.44***
Urban	.04	.07	.07	.10
Party ID		.08**	.07*	.06*
Ideology		.04	.02	.01
Knowledge		.00	−.01	.26*
Individualism			.09**	.08**
Egalitarian			−.05**	.11
Material/postmaterial			−.06	−.06
Egalitarian × Knowledge				−.04**
				(.02)

*.05 < p < .10; **p < .05; ***p < .01

the state. Both effects are strong. Age and education fail to predict attitudes toward nuclear power.

When party identification, ideology, and knowledge are added in the second equation, only party identification is large enough to be significant. Republicans are more supportive than Democrats of nuclear power. Again, political knowledge has no impact on people's attitudes toward nuclear power. The other variables remain roughly as they were in the first equation.

In the third equation, both individualism and egalitarianism influence attitudes toward nuclear power in the expected directions. Individualists support nuclear power; egalitarians oppose it. Gender and living in a county with a nuclear power reactor remain strong influences on attitudes. Party identification fades to borderline significance ($p < .06$).

Finally, we turn to the full equation. Gender still has a strong, significant impact. Whatever makes women oppose nuclear power and men support it, it is not cultural values. Controlling for all three cultural values (as well as the other variables), men and women still differ by about a half point on the five-point nuclear power scale—a substantial difference. With the addition of the egalitarianism-knowledge interaction term, however, the knowledge coefficient, which had been zero, becomes large and positive. That is, the more knowledgeable tend to support nuclear power. However, the interaction term also shows that knowledgeable people lean toward their cultural predispositions. The coefficient on the interaction term is negative, indicating that the more knowledgeable an *egalitarian* becomes, the more he or she opposes nuclear power. Again we see that knowledge and cultural values are the forces driving people's opinions. Overall, the pattern of responses to nuclear power, offshore oil development, and oil drilling in parks and government forests are essentially the same.

OVERVIEW OF FINDINGS

Let us draw this chapter to a close with an overview of the findings and an attempt to put them in perspective. We began by examining a series of demographic variables, including two—age and education—that are often identified as having the most consistent effects on people's environmental opinions. The regression equations in the last section, however, show that age and education do not have any direct effects once party, ideology, cultural values, and knowledge are included in the models. This latter set of values, therefore, explains why education and age cause environmental opinions. The young and old, and the well and poorly educated differ in their party affiliations, their ideologies, and, more important, their cultural values.

Surprisingly, the same cannot be said for race and ethnicity. Asians are far more favorably inclined toward offshore oil drilling than other groups, and Asians, blacks, and Latinos are more favorably inclined than whites toward oil drilling in parks and forests. A number of previous studies, mostly concentrating on "environmental concern," have found that blacks are no different from whites. Here we see that this finding does not extend to energy production. Whites and nonwhites do differ, once other variables are controlled. What those differences are remains to be discovered.

Gender does not seem to make much difference in attitudes toward oil development, but as previous studies have found, it strongly influences attitudes toward nuclear power. Once again, exactly what aspect of gender roles leads to this difference remains to be discovered. The greater liberalism and egalitarianism of women may explain a small part of the gender gap, but most of the gap remains unexplained.

Our last demographic variables, living in an urban area or in a county with offshore oil wells also had effects, but not ones that could have been predicted from earlier studies. Urbanites are more environmental than others on parklands, but not on other issues. The issue makes a difference—a finding that has not received a great deal of attention to date. Living near a nuclear power plant might be expected to make one anti–nuclear power; in fact, it seems to do just the opposite. Even controlling for a host of other variables, people who lived in counties with power plants favored them more strongly than people who lived elsewhere. The NIMBY syndrome completely failed to appear.

Party identification and ideology generally influenced attitudes, but they certainly did not dominate the models. When they were included with the cultural values (but not with the interaction terms), they held their own. Contrary to the findings of Ellis and Thompson in their Oregon study, party and ideology worked about as well as the cultural variables. Had Ellis and Thompson looked at the California data, instead of their Oregon data, their support for Douglas and Wildavsky's cultural theory would have been a good deal weaker.

The findings about knowledge, in contrast, generally fit with the claims made by Wildavsky and others about its limited influence. In two of the three models—offshore oil drilling and nuclear power—knowledge did not seem to have any effect until knowledge–cultural value terms were introduced.

The postmaterialism index generally failed. In two of the three models, its coefficients were statistically insignificant. Only in the equation explaining attitudes toward offshore oil development did the coefficient attain statistical significance, but in that equation it was dwarfed by the far larger coefficients for individualism, egalitarianism, knowledge, and their interactions. Simply put, Inglehart's postmaterialism theory contributes very little to our understanding of people's attitudes toward energy production.

Finally, we come to the cultural values–knowledge interactions. When these terms were introduced in the final, complete models, they dominated the results. Their coefficients—along with those of the cultural values and knowledge— were huge. With the addition of the knowledge interaction terms, Douglas and Wildavsky's cultural theory suddenly seems far more powerful. To put it another way, these equations show that cultural theory does not stand on its own. It needs the interaction terms predicted by the RAS model in order to work well. The two theories complement one another, and their combination gives us a far better understanding of public opinion on environmental issues than has anything before.

CONCLUSION

One way to evaluate a theory is to compare it to rival theories. We have already seen that the combined RAS-cultural theory predicts attitudes on energy issues better than does either Douglas and Wildavsky's cultural theory standing alone or Inglehart's postmaterialism theory. Another point of comparison is how well the theories explain change in public opinion over time. From this perspective, the combined RAS-cultural theory is also far better than its rivals.

The cultural and postmaterialism theories are basically static. Cultural theory offers no explanation of why people would change over time, except perhaps that cultural worldviews will change slowly over the course of decades and centuries as external constraints change. Postmaterialism theory predicts change in people's values, but only at the glacial pace of generational change. Neither theory has anything to say about why people might change their attitudes over relatively brief periods. Why, for example, did public sentiment turn against nuclear power? Why was a majority in favor of developing more oil resources in the 1970s, but overwhelmingly opposed ten or fifteen years later? The cultural and postmaterialism theories are silent.

The RAS model offers a fairly straightforward answer to questions about change over time. Circumstances change, political and business leaders respond, and new messages flood the news media. In the 1950s, nuclear power was claimed to be safe and so cheap that power companies would not bother to meter it. By the end of the1980s, nuclear power's claims of safety had been shattered and its cost had risen sharply. The stream of news stories reporting nuclear power plant accidents, questions about safety, and construction-cost overruns had an obvious effect on public opinion. The energy crises and the associated rise and fall of the cost of energy played important roles as well. The story with oil development is somewhat simpler. There was no gripping disaster after the 1969 Santa Barbara oil spill. Instead, the story focuses on OPEC embargoes, gas lines, and rising and falling gasoline prices.

No theory of public opinion can ignore the dynamic way in which events influence opinions. The cultural and postmaterialism theories are weak simply because they are static theories without any mechanism to explain rapid change or take current events into account. Combining the cultural and RAS theories overcomes this weakness and offers a far stronger basis for understanding the growth of environmentalism in the United States.

NOTES

1. Voter registration tapes show party registration, not party identification, but the two are closely related. Ethnicity is available because campaign consultants have developed computer programs that recognize Latino and Asian last names with a high degree of accuracy.

2. The major ideological theory that I do not investigate is the "new environmental paradigm" approach developed by Riley E. Dunlap and Kent D. Van Liere ("The New Environmental Paradigm: A Proposed Measuring Instrument and Preliminary Results," *Journal of Environmental Education* 9 [1978]: 10–19). I omit this theory for two reasons. First, it develops scales measuring people's environmental worldviews, which are proximate causes of their opinions on current environmental issues. Consider, for example, a researcher who knows that a person sees the world from a deeply pro-environmental perspective. The researcher could safely predict that the person would oppose offshore oil drilling and nuclear power. Yet these would hardly be great leaps of predictive power. By contrast, consider the researcher who knows that someone is an egalitarian—that is, the person believes in greater economic equality among all people. Why should such a theory lead to a prediction that egalitarians are opposed to nuclear power? (It does, but I will explain that later.) It follows that this theory is more powerful than environmental paradigm theory in two ways. It allows greater predictive leaps and it offers predictions in a wide range of policy areas beyond environmentalism. My second reason for omitting the new environmental paradigm scales from my research is more prosaic. I lacked the funding to ask all the survey questions that would have been useful. Finally, I should say in defense of the research on environmental paradigms that it is extremely useful for some purposes. In fact, given the funds, I would have included survey questions from that approach so that I could test them as intervening variables in a more fully specified set of statistical models.

3. John R. Zaller, *The Nature and Origins of Mass Opinion* (Cambridge: Cambridge University Press, 1992); Benjamin I. Page and Robert Y. Shapiro, *The Rational Public: Fifty Years of Trends in Americans' Policy Preferences* (Chicago: University of Chicago Press, 1992).

4. The three surveys are California Polls 8104 and 9004, and the California Offshore Oil Drilling and Energy Policy Survey. See the data appendix for full details of the surveys.

5. Iraq invaded Kuwait on August 2, 1990. The Field Poll was conducted August 17–27, 1990.

6. M. Kent Jennings and Richard G. Niemi, *The Political Character of Adolescence* (Princeton, N.J.: Princeton University Press, 1974), and *Generations and Politics: A Panel Study of Young Adults and Their Parents* (Princeton, N.J.: Princeton University Press, 1981).

7. Philip E. Converse, "The Nature of Belief Systems in Mass Publics," in *Ideology and Discontent,* ed. David Apter (New York: Free Press, 1964).

8. Converse, "Nature of Belief Systems," 211.

9. John Zaller. "Bringing Converse Back In: Modeling Information Flow in Political Campaigns," *Political Analysis* 1 (1989): 181–234; John Zaller, "Information, Values, and Opinion," *American Political Science Review* 85 (December 1991): 1215–38; John R. Zaller, *Nature and Origins*.

10. Zaller, *Nature and Origins*, 42, 44, 48–49.

11. Richard M. Scammon and Ben J. Wattenberg, *The Real Majority* (New York: Coward, McCann and Geoghegan, 1970). Others who use the distinction include William G. Mayer, *The Changing American Mind: How and Why American Public Opinion Changed between 1960 and 1988* (Ann Arbor: University of Michigan Press, 1992), and Page and Shapiro, *The Rational Public*. Other researchers have developed similar distinctions using different labels. See Byron E. Shafer and William J. M. Claggett, *The Two Majorities: The Issue Context of Modern American Politics* (Baltimore: Johns Hopkins University Press, 1995); Paul M. Sniderman with Michael Gray Hagen, *Race and Inequality: A Study in American Values* (Chatham, N.J.: Chatham House, 1985); Ronald Inglehart, *The Silent Revolution: Changing Values and Political Styles among Western Publics* (Princeton, N.J.: Princeton University Press, 1977).

12. Eric R. A. N. Smith, *How Political Activists See Offshore Oil Development: An In-depth Investigation of Attitudes on Oil Development* (Camarillo, Calif.: U.S. Minerals Management Service, 1998); Andy Caldwell, "Dealing with Facts, Figures Involving Local Oil Industry," *Santa Barbara News-Press*, 25 February 1996; Andrew LePage, "Both Sides Spouting Off on Oil Issue," *Santa Barbara News-Press*, 2 March 1996, A1.

13. This pattern is not necessarily selfishness. A good deal of research suggests that people think and act in the interests of large groups, not just themselves. However, because people understand and empathize with others who are like them, in the aggregate this looks like narrow self-interest.

14. Riley E. Dunlap and Kent D. Van Liere, "The Social Bases of Environmental Concern: A Review of Hypotheses, Explanations, and Empirical Evidence," *Public Opinion Quarterly* 44 (Summer 1980): 181–97; D. E. Morrison, K. E. Hornback, and W. K. Warner, "The Environmental Movement: Some Preliminary Observations and Predictions," in *Social Behavior, Natural Resources, and the Environment*, eds. William R. Burch, Jr., Neil H. Cheek, Jr., and Lee Taylor (New York: Harper and Row, 1972), 259–79; Ronald Inglehart, "Value Priorities and Social Change," in *Political Action: Mass Participation in Five Western Democracies* (Beverly Hills, Calif.: Sage, 1979).

15. On the World Wide Web at http://www.polsci.ucsb.edu/faculty/smith/.

16. Dunlap and Van Liere, "Social Bases of Environmental Concern." See also Robert Emmet Jones and Riley E. Dunlap, "The Social Bases of Environmental Concern: Have They Changed over Time?" *Rural Sociology* 57 (1992): 28–47.

17. David J. Webber, "Is Nuclear Power Just Another Environmental Issue? An Analysis of California Voters," *Environment and Behavior* 14 (1982): 72–83.

18. Samuel A. Stouffer, *Communism, Conformity, and Civil Liberties: A Cross-Section of the Nation Speaks Its Mind* (Garden City, N.Y.: Doubleday, 1955); Robert W. Jackman, "Prejudice, Tolerance, and Attitudes toward Ethnic Groups," *Journal of Politics* 34 (1972): 753–73; Herbert McClosky and John Zaller, *The American Ethos: Public Attitudes toward Capitalism and Democracy* (Cambridge, Mass.: Harvard University Press, 1984), chap. 8 (coauthor Dennis Chong).

19. Paul R. Abramson, *Political Attitudes in America: Formation and Change* (San Fran-

cisco: Freeman, 1983), chap. 14; James A. Davis, "Communism, Conformity, Cohorts, and Categories: American Tolerance in 1954 and 1972–73," *American Journal of Sociology* 81 (1975): 491–513; Clyde Z. Nunn, Harry J. Crockett, Jr., and J. Allen Williams, Jr., *Tolerance for Nonconformity: A National Survey of Americans' Changing Commitment to Civil Liberties* (San Francisco: Jossey-Bass, 1978). For an argument that this trend is weakening in post–World War II generations, see Thomas C. Wilson, "Cohort and Prejudice: Whites' Attitudes toward Blacks, Hispanics, Jews, and Asians," *Public Opinion Quarterly* 60 (Summer 1996): 253–74.

20. Stouffer, *Communism, Conformity, and Civil Liberties*, 90.

21. For data showing this, see Harold W. Stanley and Richard G. Niemi, *Vital Statistics on American Politics*, 5th ed. (Washington, D.C.: Congressional Quarterly, 1995), 364.

22. Mayer, *The Changing American Mind*, 176–78.

23. Inglehart, "Value Change in Industrial Societies," 1297.

24. See Van Liere and Dunlap, "Social Bases of Environmental Concern," for their well-known review of previous studies.

25. David L. George and Priscilla L. Southwell, "Opinion on the Diablo Canyon Nuclear Power Plant: The Effects of Situation and Socialization," *Social Science Quarterly* 67 (1986): 722–35; Webber, "Is Nuclear Power Just Another Environmental Issue?"

26. "Opinion Roundup: Nuclear Power," *Public Opinion*, March 1986, 26; Lawrence S. Solomon, Donald Tomaskovic-Devey, and Barbara J. Risman, "The Gender Gap and Nuclear Power: Attitudes in a Politicized Environment," *Sex Roles* 21 (1989): 401–14.

27. Rodney Fort, "The Decline of Nuclear Power in the United States: Inherent versus Economic Anti-nuclear Sentiment," in *Nuclear Power at the Crossroads*, ed. Thomas C. Lowinger and George W. Hinman (Boulder, Colo.: International Research Center for Energy, 1994), 165–89.

28. Webber, "Is Nuclear Power Just Another Environmental Issue?" 72.

29. Deborah R. Hensler and Carl P. Hensler, "Evaluating Nuclear Power: Voter Choice on the California Nuclear Energy Initiative" (Santa Monica, Calif.: Rand, 1979); James H. Kuklinski, Daniel S. Metlay, and W. D. Kay, "Citizen Knowledge and Choices on the Complex Issue of Nuclear Energy," *American Journal of Political Science* 26 (1982): 615–42; M. Maharik and Baruch Fischhoff, "Contrasting Perceptions of the Risks of Using Nuclear Energy Sources in Space," *Journal of Environmental Psychology* 13 (1993): 243–50.

30. Brian N. R. Baird, "Tolerance for Environmental Health Risks: The Influence of Knowledge, Benefits, Voluntariness, and Environmental Attitudes," *Risk Analysis* 6 (1986): 425–35; Aaron Wildavsky and Karl Dake, "Theories of Risk Perception: Who Fears What and Why," *Daedalus* 41 (1990): 41–60; Mary Douglas and Aaron Wildavsky, *Risk and Culture* (Berkeley: University of California Press, 1982).

31. Aaron Wildavsky, "Risk Perception," *Risk Analysis* 11 (1991):15.

32. Thomas A. Arcury, T. P. Johnson, and S. J. Scollay, "Ecological Worldview and Environmental Knowledge: An Examination of the New Environmental Paradigm," *Journal of Environmental Education* 17, no. 4 (1986): 35–40, and "Sex Differences in Environmental Concern and Knowledge: The Case of Acid Rain," *Sex Roles* 16 (1987): 463–72; Thomas A. Arcury and T. P. Johnson, "Public Environmental Knowledge: A Statewide Survey," *Journal of Environmental Education* 18, no. 4 (1987): 31–37; Thomas A. Arcury, "Environmental Attitude and Environmental Knowledge," *Human Organization* 49 (Winter 1990): 300–304.

33. For a similar argument, see Arcury, "Environmental Attitude and Environmental Knowledge."

34. Frederick H. Buttel and William L. Flinn, "The Structure of Support for the Environmental Movement, 1968–1970," *Rural Sociology* 39 (1974): 56–69; Frederick H. Buttel, "Age and Environmental Concern: A Multivariate Analysis," *Youth and Society* 10 (March 1979): 237–56; Paul Mohai and Ben W. Twight, "Age and Environmentalism: An Elaboration of the Buttel Model Using National Survey Evidence," *Social Science Quarterly* 68 (1987): 798–815; Julie A. Honnold, "Age and Environmental Concern," *Journal of Environmental Education* 16, no. 1 (Fall 1984): 4–9; Jones and Dunlap, "The Social Bases of Environmental Concern."

35. Zaller, *Nature and Origins*, 48.

36. Elizabeth Adell Cook, Ted G. Jelen, and Clyde Wilcox, *Between Two Absolutes* (Boulder, Colo.: Westview, 1992), 44–48.

37. These data are available on the author's Web page: www.polsci.ucsb.edu/faculty/smith/.

38. Joseph Harry, Richard Gale, and John Hendee, "Conservation: An Upper-Middle Class Social Movement," *Journal of Leisure Research* 1 (1969): 246–54; William B. Devall, "Conservation: An Upper-Middle Class Social Movement: A Replication," *Journal of Leisure Research* 2 (1970): 123–25; Marjorie Randon Hershey and David B. Hill, "Is Pollution 'A White Thing'? Racial Differences in Preadults' Attitudes," *Public Opinion Quarterly* 41 (1977–1978): 439–58; Dorceta E. Taylor, "Blacks and the Environment: Toward an Explanation of the Concern Gap between Blacks and Whites," *Environment and Behavior* 21 (1989): 175–205.

39. Paul Mohai, "Black Environmentalism," *Social Science Quarterly* 71 (December 1990): 744–65; Robert Emmet Jones and Lewis F. Carter, "Concern for the Environment among Black Americans: An Assessment of Common Assumptions," *Social Science Quarterly* 75 (September 1994): 560–79; Susan Caris Cutter, Community Concern for Pollution: Social and Environmental Influences," *Environment and Behavior* 13 (January 1981): 105–24; Jones and Dunlap, "The Social Bases of Environmental Concern."

40. Debra J. Davidson and William R. Freudenburg, "Gender and Environmental Concerns: A Review and Analysis of Available Research," *Environment and Behavior* 28 (1996): 302–39; Keith T. Poole and L. Harmon Ziegler, *Women, Public Opinion, and Politics: The Changing Political Attitudes of American Women* (New York: Longman, 1985); Robert Y. Shapiro and Harpreet Mahajan, "Gender Differences in Policy Preferences: A Summary of Trends from the 1960s to the 1980s," *Public Opinion Quarterly* 49 (1986): 42–61; Tom W. Smith, "The Polls: Gender and Attitudes toward Violence," *Public Opinion Quarterly* 48 (1984): 384–96.

41. William R. Freudenburg and Robert Gramling, *Oil in Troubled Waters: Perceptions, Politics, and the Battle over Offshore Oil Drilling* (Albany: State University of New York Press, 1994); Thomas Raymond Wellock, *Critical Masses: Opposition to Nuclear Power in California, 1958–1978* (Madison: University of Wisconsin Press, 1998).

42. Barbara Farhar-Pilgrim and William R. Freudenburg, "Nuclear Energy in Perspective: A Comparative Assessment of the Public View," in *Public Reaction to Nuclear Power: Are There Critical Masses?* ed. William R. Freudenburg and Eugene A. Rosa (Boulder, Colo.: Westview, 1984), 183–203; William L. Rankin, Stanley M. Nealey, and Barbara Desow Melber, "Overview of National Attitudes toward Nuclear Energy: A Longitudinal Analysis," in *Public Reaction to Nuclear Power,* ed. Freudenburg and Rosa, 41–67; Eugene A. Rosa and Riley E. Dunlap, "Nuclear Power: Three Decades of Public Opinion," *Public Opinion Quarterly* 58 (1994): 295–325.

43. George and Southwell, "Opinion on the Diablo Canyon Nuclear Power Plant"; Joop van der Pligt, J. Richard Eiser, and Russell Spears, "Attitudes toward Nuclear Energy: Familiarity and Salience," *Environment and Behavior* 18 (January 1986): 75–93, and "Nuclear Waste: Facts, Fears, and Attitudes," *Journal of Applied Social Psychology* 17, no. 5 (1987): 453–70.

44. The material on proximity in this section depends heavily on work I shared with Sonia Garcia, reported in Eric R. A. N. Smith and Sonia R. Garcia, "Evolving California Opinion on Offshore Oil Development," *Ocean and Coastal Management* 26 (1995): 41–56.

45. Some readers may suspect that the oil-county data are based on very small samples. The number of people living in oil-producing counties in 1981 is somewhat small (n = 73), but the 1998 data are based on the Santa Barbara/Ventura oversamples, and the numbers are quite large (n = 413). 1990 data are not mentioned because the number of Santa Barbara and Ventura residents was too small.

46. 1981 data were not used because the data set coded counties in groups and did not allow the four counties with nuclear power plants to be separated from other counties.

47. Van der Pligt et al., "Attitudes toward Nuclear Energy," and "Nuclear Waste."

48. John Lankford, "Vote Split on Goleta as City, Polls Show," *Santa Barbara News-Press*, 23 October 1990, A1. The surveys of representative cross-sections of Santa Barbara adults were conducted by Richard Hertz Consulting of Bodega Bay. The questions were: "Do you strongly support, moderately support, moderately oppose or strongly oppose oil drilling off the Santa Barbara coast?" (October 1988) and "Do you favor or oppose additional oil drilling off the Santa Barbara coast?" (October 1990).

49. If we were to assume that the questions and sampling techniques were identical, we could say that the 37 percent support in the *Santa Barbara News-Press* poll is statistically the same as the 34 percent in the California poll. However, because of different wordings of questions and the fact that different firms did the polling, such precise claims are shaky.

50. This survey was conducted well before the Mobil Oil Corporation's announcement of its "Clearview" proposal to remove an offshore oil platform and build a slant-drilling site in Isla Vista, a student housing neighborhood at the edge of the UCSB campus. The Clearview proposal, therefore, did not influence this survey.

51. In both questions, some students checked the "Don't Know/No Opinion" box. Putting aside the missing data, the sample sizes were 224 for the local development question and 221 for the remote development question.

52. Freudenburg and Gramling, *Oil in Troubled Waters*, 35.

53. Kenneth R. Tremblay, Jr., and Riley E. Dunlap, "Rural-Urban Residence and Concern with Environmental Quality: A Replication and Extension," *Rural Sociology* 43 (1978): 474–91.

54. Van Liere and Dunlap, "Social Bases of Environmental Concern"; Jones and Dunlap, "Social Bases of Environmental Concern."

55. David Kowalewski, "Environmental Attitudes in Town and Country: A Community Survey," *Environmental Politics* 3 (Summer 1994): 295–311.

56. William R. Freudenburg, "Rural-Urban Differences in Environmental Concern: A Closer Look," *Rural Sociology* 61 (May 1991): 167–98.

57. James McEvoy, "The American Concern with the Environment," in W. R. Burch et al., eds., *Social Behavior, Natural Resources, and the Environment* (New York: Harper and Row, 1972).

58. U.S. Bureau of the Census, *1990 Census of Population. General Population Characteristics, California* (Washington, D.C.: Bureau of the Census, 1992), app. 2, 937–44.

59. Robert C. Paehlke, *Environmentalism and the Future of Progressive Politics* (New Haven, Conn.: Yale University Press, 1989), 95, 177.

60. Paehlke, *Environmentalism and the Future of Progressive Politics,* chap. 7.

61. Scammon and Wattenberg, *The Real Majority;* Mayer, *The Changing American Mind;* Page and Shapiro, *The Rational Public;* Shafer and Claggett, *The Two Majorities;* Sniderman with Hagen, *Race and Inequality.*

62. Joseph P. Kalt and Mark A. Zupan, "Further Evidence on Capture and Ideology in the Economic Theory of Politics," *American Economic Review* 74 (1984): 279–300; Sheldon Kamieniecki, "Political Parties and Environmental Policy," in *Environmental Politics and Policy,* ed. James P. Lester, 2d ed. (Durham, N.C.: Duke University Press, 1995), 146–67.

63. See Bruce E. Keith, David B. Magleby, Candice J. Nelson, Elizabeth Orr, Mark West-lye, and Raymond E. Wolfinger, *The Myth of the Independent Voter* (Berkeley: University of California Press, 1992) for their discussion of both reasons.

64. Abramson, *Political Attitudes in America,* chaps. 5, 6.

65. Political awareness scores of zero and one were combined because of the small number of respondents with those low scores. Unfortunately, these information questions are available only in the 1998 data, so I could not investigate the RAS model using 1981 or 1990 data. I tried using education in place of the information scale, but, as Zaller observes, education does not work very well.

66. The wording of the ideology questions changed slightly between 1990 and 1998, but it is unlikely that the minor change had any effect on the distance between liberals and conservatives. The 1981–1990 wording asked respondents whether they were "strongly [liberal/conservative] or just moderately [liberal/conservative]," while the 1998 wording asked whether they were "strong or not very strong" liberals or conservatives. The larger number of categories in the 1998 question is caused by a finer distinction among moderates in a different branch of the question.

67. Eric R. A. N. Smith, *The Unchanging American Voter* (Berkeley: University of California Press, 1989); Michael X. Delli Carpini and Scott Keeter, *What Americans Know about Politics and Why It Matters* (New Haven, Conn.: Yale University Press, 1996).

68. Edward G. Benson, "Three Words," *Public Opinion Quarterly* 4 (1940): 130–34; Philip E. Converse, "Nature of Mass Belief Systems"; John C. Pierce, "Party Identification and the Changing Role of Ideology in American Politics," *Midwest Journal of Political Science* 14 (1970): 25–42; Norman R. Luttbeg and Michael M. Gant, "The Failure of Liberal/Conservative Ideology as a Cognitive Structure," *Public Opinion Quarterly* 49 (1985): 80–93.

69. Douglas and Wildavsky, *Risk and Culture;* Mary Douglas, *Risk and Blame: Essays in Cultural Theory* (London: Routledge, 1992); Aaron Wildavsky, *The Rise of Radical Egalitarianism* (Washington, D.C.: American University Press, 1991); Wildavsky and Dake, "Theories of Risk Perception." See also Richard J. Ellis and Fred Thompson, "Culture and the Environment in the Pacific Northwest," *American Political Science Review* 91 (December 1997): 885–97.

70. Claire Marris, Ian H. Langford, and Timothy O'Riordan, "A Quantitative Test of the Cultural Theory of Risk Perceptions: Comparison with the Psychometric Paradigm," *Risk Analysis* 18 (1998): 635–47.

71. The reliability (Cronbach's alpha) of the egalitarianism index is 0.72; the reliability of the individualism index is 0.54.

72. Ellis and Thompson, "Culture and the Environment in the Pacific Northwest."

73. A number of other scholars treat individualism and egalitarianism as basic values in their studies without relying on cultural theory to justify them. See, for example, Stanley Feldman, "Measuring Issue Preferences: The Problem of Response Instability," *Political Analysis* 1 (1989): 25–60; Stanley Feldman and John Zaller, "The Political Culture of Ambivalence: Ideological Responses to the Welfare State," *American Journal of Political Science* 36 (1992): 268–307; Sniderman with Hagen, *Race and Inequality*.

74. Some readers might also expect the individualism and egalitarianism indexes to correlate more strongly than the ideology question with the opinions on energy issues. This might occur because statistical theory tells us that a series of questions about a subject generally yields a better, more reliable measure than does a single question, because with a series of questions, small random errors in the answers tend to cancel one another out. This is the logic explaining why, for example, teachers use many arithmetic questions, rather than a few, when making up tests to grade students. More questions should yield more reliable answers, all things being equal. Furthermore, reliable measures tend to correlate with other variables more strongly than do unreliable measures. In other words, if ideology does cause environmentalism, then improving the reliability of the ideology measure should reveal a strength of correlation that had been hidden (or in technical terms, "attenuated") by measurement error.

Although this argument sounds good in theory, it does not work out in practice. The ideology measure has proven to be fairly reliable in other surveys, and the individualism and egalitarianism indexes are less reliable despite having been built with three items each.

75. Wildavsky, *Rise of Radical Egalitarianism*.

76. The data are shown on the author's Web page: www.polsci.ucsb.edu/faculty/smith/.

77. Ronald Inglehart, "Value Priorities"; Paul R. Abramson and Ronald Inglehart, *Value Change in Global Perspective* (Ann Arbor: University of Michigan Press, 1995); Inglehart, *Silent Revolution*.

78. See Abraham Maslow, *Motivation and Personality* (New York: Harper and Row, 1954).

79. James A. Davis, "Review Essay on Paul R. Abramson and Ronald Inglehart, *Value Change in Global Perspective*," *Public Opinion Quarterly* 60 (1996): 322–31.

80. Van Liere and Dunlap, "Social Bases of Environmental Concern"; Jones and Dunlap, "Social Bases of Environmental Concern."

81. Alternative coding schemes yield no substantive difference in the findings. I estimated models omitting respondents with no opinions, and models using the same dichotomous, agree-disagree scoring used in the figures. Nothing of consequence changed.

82. The choice to exclude whites and use them as the baseline category is statistically arbitrary. Any category could have been excluded without materially changing the results. The coefficient of 0.37 indicates that blacks score 0.37 higher than whites in their support for drilling. Had I excluded blacks instead of whites, the coefficient for whites would have been − 0.37, indicating that whites score 0.37 lower than blacks.

83. On how to interpret R-square, see Michael S. Lewis-Beck and Andrew Skalaban, "The R-Squared: Some Straight Talk," in James A. Stimson, *Political Analysis* 2 (1990): 153–72; Christopher H. Achen, "What Does 'Explained Variance' Explain? Reply," *Political Analysis* 2 (1990): 173–84; and Gary King, "Stochastic Variation: A Comment on Lewis-Beck and Skalaban's 'The R-Squared,' " *Political Analysis* 2 (1990): 185–200.

84. See G. S. Maddala, *Introduction to Econometrics* (New York: Macmillan, 1988) or any other standard econometrics textbook for an explanation of multicollinearity.

85. Recall that the Cronbach alpha for the individualism index is 0.54.

86. More precisely, in figure 2 the height of the graph, y, is equal to 0.335* (knowledge) + 0.234* (egalitarianism) − 0.063* (Knowledge × Egalitarianism). These are the coefficients from equation 4 extended to one more digit for accuracy.

6

Implications for the Future

We now have a collection of findings about public opinion and energy. The time has come to take a step back and consider their implications for the future. There is more to this than simply projecting opinion trends into the future. The real questions here are: What are future energy-related conditions and events likely to be? How will people respond to those changing conditions and events? How will politicians respond to the public's demands? To answer those questions, we must consider the energy situation that we will face in the future, and look back at the past to see how people are likely to respond to coming changes in energy supplies.

THE LIMITED SUPPLY OF OIL

The starting point for any discussion of future energy policy is the fact that the world supply of petroleum is finite. It will not last forever. In all likelihood, our grandchildren will live in times of substantially reduced oil consumption. More immediately, diminishing oil supplies will probably begin to drive up prices within the next fifteen to twenty-five years.

Economists typically respond to such dire warnings by claiming that the market will solve the problem. Alternative fuels will become available. Businesses will invest in more energy-efficient equipment. People will adjust their buying habits and learn to accept mass transit. Realistically, government regulation will play an important role as well. If people do not stop buying SUVs because of the high cost of gasoline, Congress may legislate them out of existence.

That may be an economist's notion of a solution, but it is hardly one that will please many people. Even relatively trivial steps like banning high-gas-mileage cars, such as SUVs, would cause the auto industry and their potential customers to scream at Washington. We have already seen the incredible anger produced by higher gas prices. Higher energy prices could also bring either more rounds of inflation or recessions, or even both simultaneously as they have in the past (recall the "stagflation" of the 1970s). In short, the economists' solution could be ugly.

Some critics suggest that economists may fail to grasp the problem fully. Economists, along with just about everybody else, assume that scientists will be able to solve all our problems for us. If scientists can put a man on the moon and eliminate such devastating diseases as smallpox and polio, surely they can overcome energy shortages—or at least that is what most of us think. But could we be wrong? After all, petroleum has been one of the primary energy sources for almost the entire industrial age. What if the wonderful new alternative-energy sources are simply not able to produce enough energy to replace petroleum? It is now over a quarter-century since the first energy crisis struck, and all the geothermal, solar, and wind energy together still account for only about one-half of 1 percent of all energy produced in the United States (and we consume far more than we produce). What if efforts to develop alternative fuels fail to produce sufficient quantities at affordable prices? What if efforts to conserve energy fall short of their goals?

The answer, of course, is that we still have coal and nuclear power. We have so much coal that if we were to substitute coal for petroleum, we would have no energy problem. Yet this is obviously not a perfect solution. Coal produces not only electricity but air pollution, acid rain, and possibly global warming. Nuclear power produces radioactive and toxic waste and a real, if very small, risk of nuclear accidents. As a consequence, we can safely say that the worst scenarios of running out of energy will never happen. Instead, we will face less dramatic, but still painful decisions about using coal and nuclear power and about conserving energy.

A second aspect of the energy situation in the coming century is that from time to time OPEC will flex its muscles and drive up oil prices. It has successfully done this several times already. Although it is often forgotten, oil boycotts were attempted even before the 1973–1974 energy crisis. In fact, when I wrote the first chapter of this book, in 1999, oil prices were at an all-time low. I offered this same warning about the potential for OPEC to drive up prices. Then in the early months of 2000, OPEC cut back on production and prices shot up to over thirty-five dollars a barrel again. What might have been a nice prediction turned into an inconvenient job of rewriting. The high prices of 2000 and 2001 will probably drop back to a more moderate range within a year or so, but far more certainly, there will be more price spikes in the future.

Political scientists and economists tell us that OPEC is not likely to be able to maintain tight control of the oil market in the long run. Each member nation and every oil company has an incentive to cheat. If OPEC manages to set a twenty-eight-dollar-a-barrel price, those who offer oil at a slightly lower price will be able to increase their sales enough to gain a larger profit. Because OPEC has a large number of members, some of whom have fought wars against one another in the not-too-distant past, OPEC is inherently unstable. Members will cheat and OPEC's attempts to keep prices high all the time will fail. Nevertheless, it can still muster the will to drive up prices occasionally. Those price spikes can cause havoc both among the consumer nations' economies and with their voters.

In sum, the future is likely to bring declining oil supplies, rising prices, and occasional OPEC-induced price jumps. Given this situation, we can ask, how are politicians and the public likely to respond?

POLITICIANS AND PUBLIC OPINION

In a democratic system such as the United States, public opinion is a powerful force. It is not the only one, of course. When making policy decisions, politicians consider the views of lobbyists, interest groups, their personal staffs, and even other politicians. They often know a good deal more about any given policy debate than the public, and that knowledge influences their decisions as well. Finally, their personal preferences play a role in their decisions. Yet out of all these considerations, public opinion weighs especially heavily because politicians who defy the public run the risk of losing reelection.

Critics on both the Left and Right often charge that powerful, special interests dictate what politicians will decide. Congress passes laws that big business demands and ignores what the people want. In particular, multinational oil companies have often been targeted as corporate evildoers by people who dislike laws coming out of Congress. Yet the history of energy policy has largely been one of the public getting what it wants. This has not been universally true—the oil industry has won some battles—but the oil industry's victories have generally come when the public was not paying attention to energy issues. When the public has focused on energy policy and gotten angry, it has usually gotten what it wanted—at least insofar as politicians could give it.

In the 1950s, President Dwight Eisenhower imposed an oil-import quota that resulted in higher prices in the United States than in the rest of the world. This might seem like an oil industry victory over public opinion, but it was falling prices in the world market that caused the gap. In the United States, gasoline and heating oil prices remained steady, and the public largely ignored them. The oil companies were winning, but the public did not object. Rising prices cause public anger; stable prices cause public indifference.

In the late 1960s and early 1970s, the price of oil on the world market rose, along with the general rate of inflation. This time the public cared. President Richard Nixon responded with wage and price controls to fight inflation. Although the oil companies, along with most other big businesses, opposed the move, the president gave the public what he thought it wanted. The wage and price controls actually made the energy situation worse. Subsidizing gasoline prices encouraged consumption, which made supplies even tighter. But the public demanded low gasoline prices and Nixon wanted to please the voters.

When prices shot up during the twin oil crises of the 1970s, the public demanded action to increase supplies and cut prices. It got action. President Jimmy Carter created the Energy Department and attempted to push a wide range of energy development and conservation programs through Congress. Not all of them passed, but some did—opening new oil fields, easing regulatory burdens on energy producers, creating a synthetic-fuels market, and pumping money into research. Yet the public simultaneously turned away from nuclear power, especially following the Three Mile Island accident. There, too, the public got what it wanted. Congress and state governments passed laws imposing more and more costly regulations and licensing hurdles on nuclear power plants. The utility companies and makers of nuclear power equipment, despite their wealth and political muscle, did not win. The nuclear industry began to die.

Finally, as the gas lines and energy shortages faded into memory in the 1980s and 1990s, the public turned against offshore oil drilling. As gas prices fell, people saw more value in the beauty of the coasts than in vast oil reserves along the California coast or the suspected oil reserves along the East Coast. The oil was there, it was attainable, but the public did not want the oil companies to get it. Again, despite the political power of the oil industry, the oil fields were denied to them.

The general point is that on high-profile issues, politicians mostly do what the public wants. They do not always bow to public wishes, but they usually do because to do otherwise is to risk defeat on election day. Oil and utility companies can make substantial campaign donations—and having lots of campaign cash certainly helps win reelection—but even lavish campaign spending cannot undo the damage that comes with unpopular positions on emotional issues.

Of course, many energy policy disputes get so little public attention that politicians can do whatever they want. Few people really care whether research on alternative fuels is adequately funded or whether insulation standards for new homes are changed. On these low-profile issues, the huge energy corporations can be very influential. Yet people do care when gasoline or electricity prices rise, when utility companies want to build nuclear power plants, and when oil companies want to drill for oil along the nation's coasts. On these sorts of issues the voters speak loudly and politicians obey.

The future will undoubtedly bring more situations in which people will get angry and politicians will scramble to do their bidding. Declining oil supplies, rising prices, and occasional OPEC-induced price spikes will upset people and put pressure on politicians to find painless solutions for the problems. The question now is what sort of solutions people will demand.

THE PUBLIC RESPONSE

How will the public respond to the coming hikes in the cost of gasoline and other sources of energy? We can make several predictions.

We can begin by predicting that people will continue to demand low, stable prices for gasoline and other forms of energy, and more important, subsidies for energy if prices rise too quickly. This is a basic aspect of the energy-crisis cycle described in chapter 2. We have seen this behavior in every energy crisis and we should expect to see it again. We have seen it with gasoline prices, heating oil prices, and electricity rates. When prices jump, people get angry and demand that the government do something about it. The usual solutions are subsidies of some kind for energy use—for example, cutting gasoline taxes, subsidizing home-heating-fuel costs for the elderly and low-income residents of the snow belt, and capping electricity rates so that consumers do not pay the full cost.

Economists shudder at the idea of subsidizing the use of an essential product when supplies are short. Economically it makes no sense because subsidies encourage consumption and cause supply problems to worsen. Politically, of course, it makes perfect sense because voters are angry, and cutting gasoline taxes or otherwise reducing energy costs pays great dividends on election day. Poor economics makes good politics.

Voters generally fail to recognize the problem with subsidizing energy use. One might think that the idea that supply and demand cause prices is well known. Indeed, a majority of Americans grasp the basic idea, although understanding even the basics is far from universal. Yet the arguments about subsidizing energy prices are complicated because there are more considerations than the simple economics of supply and demand. Energy prices have moved up rapidly in the past because of OPEC actions, Middle Eastern wars, the failure of utility companies to estimate future demand and build enough power plants, and misguided energy deregulation plans. In addition, another aspect of the energy-crisis cycle is that every time that oil prices have gone up quickly, political leaders have accused the oil companies or power companies of price gouging. So the public saw foreign "bad guys" and greedy corporate executives playing prominent roles in our energy problems.

The evidence came out clearly in public opinion polls discussed in chapter 2. Since the first energy crisis began in 1973, oil companies have been among the least trusted organizations in the United States. Moreover, during both energy crises of the 1970s, about 70 percent of the public told pollsters that the energy shortages were not real. They believed oil companies faked the crises in order to make profits. The same pattern is playing out in the current energy crisis.

What follows is that if the shortages are not real, the economics of supply and demand are not relevant. As the public sees it, the solution to political or corporate wrongdoing should be positive government action. If bad guys fix prices, the government should step in and stabilize them. Allowing energy prices to rise without limit is not free-market economics; it is allowing the bad guys to win. Looking to the future, we have no reason to expect this pattern not to repeat itself. When prices rise, the public will demand subsidies, and the energy situation will worsen.

A second prediction we can offer is that the public will not respond with carefully reasoned or well-informed policy preferences. We have seen in chapter 4 that the public is poorly informed and often confused about policy details. Most people have general pro-environment or pro-development sympathies, but translating those sympathies into preferences on concrete policy questions is difficult for them. That is why most people, regardless of whether they generally favor conservation or energy development (or both), hold a mix of opinions that seem to come from opposing political camps. A March 2001 Gallup poll, for example, found that 56 percent of the American public wanted to emphasize conservation, while only 33 percent wanted to emphasize energy production, to deal with our energy problems. Yet in the same poll, 53 percent wanted to give tax breaks to provide incentives for more oil and gas drilling in the United States.[1] That is, at least some of those respondents rejected the idea of emphasizing oil development, but wanted to create incentives for more oil development anyway. Although most public opinion surveys show the public leaning in a strongly pro-environmental direction, there are also results showing pro-development preferences. Those mixed signals provide politicians with ammunition for their disputes. Each side can point to polls supporting its policy preferences.

Global warming is an especially important example of a policy debate about which the public is poorly informed. Although a majority of people recognize the term and seem to understand that it refers to the fact that the earth is getting warmer, they are confused about possible causes. A large portion of the population incorrectly believes that the causes of pollution and global warming are the same. Moreover, the public still does not express much interest in global warming. The problem does not even rank very high among environmental concerns.[2] Moreover, at this point, it seems highly unlikely that many people recognize the possible connections between energy policy and global warming. The lack of information and of interest is a major barrier to environmentalists who hope to push the United States into action.

A related prediction we can offer is that the "carbon tax" will continue to fail to overcome the public's resistance to complicated, confusing proposals. The basic idea of the carbon tax is that fossil fuels should be taxed in proportion to the amount of greenhouse gases they produce. Burning coal produces a large amount of carbon dioxide and so it would be taxed heavily. Natural gas produces relatively less carbon dioxide for the amount of energy it produces, so it would be taxed at a lower rate. Hydroelectric, solar, and wind power yield no greenhouse gases and would not be taxed at all. The argument for the carbon tax is that we should use the tax system to discourage energy sources that cause global warming and encourage clean energy sources. Unfortunately, it is a complicated argument. The tax incentive structure is not simple. Many, probably most, people do not understand the concept of a greenhouse gas. Although a consensus has formed in the scientific community that global warming is occurring, whether using fossil fuels causes it is still publicly disputed. Persuading the public to accept any tax is tough. Persuading the public to accept a tax that is supported by a complicated argument and that has no immediate benefit is a daunting challenge.

A fourth prediction we can offer is that offshore oil drilling will return to public favor. In the past, high fuel prices and a high rate of inflation pushed people into supporting drilling for oil both offshore and in parks and national forests. There is no reason to believe that when prices rise again, the pattern will not be repeated. Moreover, the argument here is a simple one: If we produce more oil and gas, the price will go down.

We have actually seen the beginning of this pattern twice in the last few years. In 1996 and again in 2000, oil prices unexpectedly rose. Both times, angry motorists complained about the cost of gasoline. Both times, politicians responded by cutting gas taxes and talking about releasing oil from the nation's strategic petroleum reserve. Another aspect of the situation, however, was that oil industry advocates began publicly calling for opening up California's coast and the Arctic National Wildlife Refuge to more oil drilling.[3] Efforts to sway public opinion were launched.

Cutting gasoline taxes is a quick, direct way to ease consumers' pain at the gas pump. It shows voters that politicians care. Opening up new oil fields for exploration and development is a slow way to influence gasoline prices. The oil must be located, drilling operations must be set up, transportation must be arranged. The entire process from opening an oil field to output from a refinery takes years, perhaps as much as a decade in the case of oil from Alaska. As a response to a short-term hike in gas prices, it does not have much political appeal. As a response to a long-term trend of rising prices, however, it does have appeal. When months of rising prices again begin to stretch out into years (as they did in the 1973–1974 and 1979–1980 energy crises), long-term solutions such as opening up the Arctic National Wildlife Refuge and the East and West Coasts to oil development will begin to seem more attractive both to voters and to politicians. President George W. Bush may not win his fight to open up the Arctic National Wildlife Refuge during his presidency, but when the world supply of oil begins to decline and prices start rising steadily, appeals to drill more Alaskan oil will probably prevail.

The people most likely to accept the arguments favoring drilling will be those who are predisposed toward development policies—Republicans, conservatives, anti-egalitarians, and individualists. In the years since the last energy crisis, pro–oil development arguments have been largely absent from the nation's news media, and consequently, these people have become more pro-environmental. As the arguments reappear, however, these people can be expected to move back toward more development friendly views. In particular, President Bush's high-profile leadership on Arctic oil drilling should have a substantial influence on Republicans and conservatives.

People who are predisposed toward pro-environmental views—Democrats, liberals, egalitarians, and anti-individualists—will also be affected. They, too, should move in a pro-drilling direction, although not as much as people who are predisposed toward that view. Some, especially those with only moderate amounts of political knowledge, will be persuaded that more drilling is good. Even though they may lean toward environmentalism, some will not recognize that environmentalist leaders oppose drilling for oil along the California coast and in the Alaskan wilderness. In addition, inflation and high gasoline prices will influence them, just as they do everyone else.

The likely movement of public opinion toward favoring energy development highlights a critical failure on the part of the groups opposing offshore oil drilling along the California coast. The moratorium on new offshore oil leases was established by executive order. Presidents Bush and Bill Clinton instructed their Department of the Interior secretaries not to lease any more tracts for oil development. Yet with the stroke of a pen, that moratorium could be ended. What a president can give with an executive order, a president can take away just as easily. Environmentalists have attempted to put into legislation a ban on offshore oil development, but Congress has never passed one. If environmentalists were successful in putting the moratorium into law, they would find it far easier to defend if public opinion turned against them. A few well-placed members of Congress on key committees could kill an attempt to end the moratorium. For that matter, a single senator willing to filibuster to protect the California coast might be able to do it. As matters stand now, however, pro-oil interests have managed to block the environmentalists' efforts. In the coming years, that victory may well open the door for more oil drilling along the California coast.

A fifth prediction we can make is that nuclear power will have a much longer road to recovery in the public's eyes than will oil or coal. Nuclear power has been thoroughly tarnished both by the Three Mile Island accident and by antinuclear advocates. In national polls it is far less popular than oil. Moreover, the prospects of nuclear power are hampered by the fact that the connection between nuclear power and electricity rates is not as close as that between oil supplies and gasoline prices. We only get gasoline from petroleum, but we can get electricity from petroleum, natural gas, coal, hydroelectric dams, solar power, wind power, and geothermal power, as well as from nuclear power plants. Technological advances may eventually give us electric cars that have the same power and performance as gasoline-powered cars, but right now people who drive depend on oil, and they know it. People who use electricity, however, have a wide range of choices about how to produce it. Spurning nuclear power is easy.

Eventually nuclear power may return to public favor. We should remember that the inflation rate does influence support for nuclear power. As inflation and energy prices increase, so does support for nuclear power. Consequently, as the world petroleum supply diminishes, nuclear power should again seem more desirable. Another reason why nuclear power may become more popular is that it does not produce any greenhouse gases, unlike fossil fuels. If global warming becomes an increasingly serious problem, as most climate scientists believe likely, nuclear power may begin to seem more attractive than coal. Of course, a mix of alternative fuels and conservation could make nuclear power unnecessary, but there is little in the history of the last thirty years to suggest this is likely. Alternative fuels still provide only about one-half of 1 percent of all energy consumed in the United States, and despite the energy conservation efforts following the energy crises of the 1970s, Americans have returned to record-breaking levels of energy consumption per capita.[4] If global warming continues and science does not give us any major breakthroughs, nuclear power will look better and better to many Americans.

A final prediction we can offer is that by the middle of the century, the choice of energy solutions is likely to come down to coal and nuclear power. This prediction follows from three considerations. First, the evidence suggests that about half the world petroleum supply has been consumed. We will not be able to depend on petroleum as our major energy source for the rest of the century. We will therefore need to supplement it from additional sources or use less energy. Second, the public has shown little inclination to accept conservation and the lifestyle changes that accompany it as a serious option. People are happy to conserve energy so long as it is reasonably convenient and does not restrict their freedom of choice in any important way. For example, aluminum and glass recycling rates have risen enormously since the 1970s because modern programs have made recycling easy. Proposals to conserve energy in ways that people find inconvenient or unduly restrictive, however, have never been popular. Automobiles are the classic example. Here we can look at small, fuel-efficient cars and the fifty-five-mile-per-hour speed limit as examples. Neither one was ever popular. When SUVs arrived on the market, the public loved them, despite their conspicuously low gas mileage. Third, the slow pace of alternative energy research and development does not bode well for the future. Alternative energy has always seemed promising—and, in truth, it still is. Yet the glowing reports about progress on alternative energy sources have too often overlooked the problems that alternative energy faces. One problem is that wind, solar, and geothermal power are still in their infancy. Moreover, as they grow they will surely become more controversial. Public opinion surveys already show that Californians regard offshore oil platforms as ugly. Yet there are fewer than two dozen platforms along the California coast. How will the public respond to tens of thousands of windmills covering hillsides and valleys? There has been little protest so far, but if wind power is to help the U.S. energy situation, we will need a staggering number of windmills. Aside from aesthetic issues, there are also environmental objections. Windmills kill birds. In fact, the National Audubon Society attempted to block construction of a major wind farm because of the threat it posed to endangered California condors.[5] There will no doubt be aesthetic objections to solar farms if solar panels become economically competitive with other energy sources. How much of our countryside are we willing to carpet with solar panels, and what endangered species live in those areas?

As for the more conventional source of renewable energy, hydroelectric power dams, there have long been environmental opponents organized against them. Although the recent power shortages have led some people to call for building more dams, there are also groups working to prevent dams from being constructed and in some cases to remove existing dams.[6] The point of these observations is not that alternative energy sources and hydroelectric dams are bad, but that as the technologies develop and become more widely used, they will become more controversial at the same time. Alternative energy offers hope for the future, but it is no magic bullet that will solve all our energy problems.

In the next few decades, the diminishing world oil supply and problems with global warming will bring energy policy to the forefront of American politics. Unless scientists solve our problems with stunning breakthroughs in alternative energy or Americans suddenly decide to reduce their energy consumption, we face a range of unpleasant choices. The rising cost of energy will surely damage the U.S. and world economies. The diminishing energy supply will force us to rely more on coal or nuclear energy. Congress will no doubt pass more and more measures forcing people to conserve energy. None of these choices will be popular, but the earlier we address our energy problems, the better off we will be in the long run.

NOTES

1. Lydia Saad, "Americans Mostly 'Green' in the Energy vs. Environment Debate," Gallup Poll Release, www.gallup.com, 16 March 2001.

2. Wendy W. Simmons, "Despite Dire Predictions of Global Warming, Americans Have Other Priorities," Gallup Poll Release, www.gallup.com, 20 February 2001.

3. Frank H. Murkowski, "Let Alaskan Oil Help the State, Nation," *Los Angeles Times*, 17 February 2000, B9; George Reisman, "Government's Oil Prices," *Orange County Register*, 2 April 2000, 2; Eric Slater, "In Race, Rising Gas Prices Are Make-or-Brake Issue," *Los Angeles Times*, 24 June 2000, A1.

4. U.S. Energy Information Administration, *Annual Energy Review 1997* (Washington, D.C.: Government Printing Office, 1998), 9, 13.

5. Elizabeth Shogren, "Wind Farm Called Threat to Condors," *Los Angeles Times*, 14 September 1999, A3.

6. Leon D. Keith, "Pressure Building for More Dams," *Santa Barbara News-Press*, 15 April 2000, 4; Kim Murphy, "Between a Rock and a Hard Place: Where Dams, Salmon Meet," *Los Angeles Times*, 13 June 2000, A5; Keith D. Warner, "Extinction Is Forever, Dams Are Not," *Los Angeles Times*, 30 June 2000, B8.

Statistical Methods Appendix

This book uses a number of statistical methods to describe and analyze public opinion data. In this appendix, I present a brief description of how to interpret those statistics for readers who do not have a background in statistics. None of my descriptions is detailed, and none can substitute for a real statistics course. I hope only to explain the basic ideas so that readers can follow the discussion.

SAMPLING

Almost all the data presented in this book are based on public opinion surveys. Typically, a sample of about a thousand respondents is used to make claims about the entire population. How are such claims justified?

The best, most representative type of sample is what statisticians call a *simple random sample*. To draw such a sample, one would need a complete list of people in the population (for example, the United States). People would be randomly selected from the list, perhaps by writing everyone's name on separate pieces of paper and thoroughly mixing the pieces of paper in a huge bowl. If the bowl were mixed well enough to give everyone an equal chance of selection, we would have the perfect random sample.

No such complete list exists, but pollsters have developed methods to get around this. Using census data and sophisticated sampling techniques, pollsters draw random samples of regions and then neighborhoods within the regions. An exact list of households (if the survey is conducted in person by interviewers in the respondents' homes) or telephone numbers (if the survey is conducted by telephone) is needed, but only at the end of the sampling process. Just as in a simple random sample, everyone has an equal chance of being selected for the survey. The result is a sample representative of the entire population.

To provide an accurate description of the population, a sample must be not only random but large enough. It often surprises people that the size of the population does not matter, only the size of the sample. A survey of one thousand respondents will work just as well to estimate public opinion in the United States (population 285 million) as in Iowa City, Iowa (population sixty thousand). To see why this is so, try flipping a coin a number of times. After the first ten tosses, the percentage of heads may not be close to 50 percent. But after a hundred tosses, the percentage of heads will be much closer to 50 percent, and after a thousand tosses, the percentage will be very close. In fact, the chances are 95 percent that the number will fall between 46.9 percent and 53.1 percent. The 3.1 percent variation from 50 percent is called the *sampling error*.

Just as the sampling error decreases as you toss the coin an increasing number of times, the sampling error gets smaller as the sample size grows in a survey. In other words, the bigger the sample, the more accurate the result—no matter what the total population size. The sampling error on a sample of five hundred is 4.4 percent. With a thousand respondents, it falls to 3.1 percent. With a sample of fifteen hundred, it falls to 2.5 percent. Typical commercial polls use samples of about a thousand people.

STATISTICAL SIGNIFICANCE

Because samples are used to estimate opinion, there is always a chance that the sample is badly skewed. An old adage from survey research warns that some day a pollster will conduct a survey and randomly select one thousand left-handed Lithuanians. There are, after all, undoubtedly a thousand left-handed Lithuanians living in the United States, and therefore it is possible, although extremely unlikely, that a survey would choose that sample.

How do we know that random samples are representative of the entire population? A powerful result from statistical theory shows us how to estimate the accuracy of a sample. The "central limit theorem" tells us that if we use a random sample, we can calculate a *confidence interval*, which allows us to make statements such as: "There is a 95 percent probability that the true number is 50 percent plus or minus 3.1 percent." That is, after we estimate a mean or some other statistic using a sample, we can figure out how close the true population mean is to the mean from our sample. By widespread convention, scientists use the 95 percent confidence level. Of course, there is always a possibility that we did draw a sample of one thousand left-handed Lithuanians, but if we are 95 percent certain about our answer, we have strong enough evidence to draw conclusions safely.

Another result of the central limit theorem is that we can calculate a *p-value*, or *statistical significance level*, for a wide variety of statistical measures—from difference-in-means tests to correlations to regression coefficients (discussed below). The p-value is the probability that a difference (for example, the difference in support for nuclear power between men and women) is the product of random sampling fluctuation. When we say that a difference is statistically significant at p < .05, we mean that that there is less than a 5 percent chance that the difference is the result of random sampling error. To put it the other way, there is greater than a 95 percent probability that the difference is real. Again, the conventional standard used by most scientists is 95 percent confidence.

REGRESSION ANALYSIS

"Ordinary least-squares regression analysis" is probably the most widely used method of assessing the impact of independent, causal variables on a dependent variable. In its simplest form, regression offers a precise estimate of the causal effect of one variable on another by calculating a line that best describes the relationship. For example, figure A1 presents a small, hypothetical sample of people's educations and incomes. Each point in the graph represents a single person. The point in the upper right, for instance, represents a person with twenty-three years of school (a Ph.D.) who makes $56,500 a year. The regression line in the figure summarizes the relationship. As education increases, so does income.

Figure A1   A Hypothetical Regression Line

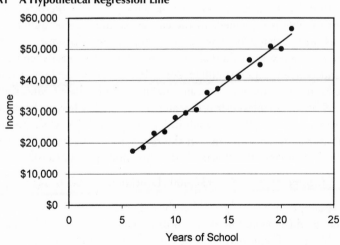

The equation for the line is:

$$\text{Income} = \$2{,}000 + \$2{,}500 \ (\text{Education})$$

The first regression coefficient, $2,500, is the *constant,* or *intercept.* The second coefficient, $2,500, is the *slope coefficient,* or *b,* for education. The intercept is the point on the y-axis the line strikes when income is zero. The slope coefficient predicts that for each additional year of education, the expected increase in income is two thousand dollars. So a high school graduate would be expected to earn $2,000 + $2,500 (12 years) = $32,000 per year. The regression model estimates the coefficients by finding a line that minimizes the sum of the squared errors between each observation and the line. In that sense, it is the line that best describes the set of observations.

In addition to calculating the coefficients for the regression line, the method calculates a set of *standard errors,* which allow users to construct "confidence intervals." A coefficient is statistically significant at the .05 level if it is large enough to give less than a 5 percent chance that the coefficient differs from zero because of random sampling fluctuation. When someone writes that a coefficient is statistically significant, he or she is saying that the evidence is overwhelming that the coefficient really is large and that the independent causal variable really does have an effect on the dependent variable. In the education and income example, the education coefficient is statistically significant at $p <$ .01, so we can be very confident that education causes income.

The regression method also calculates a number known as the "adjusted R^2" (that is, R-squared) that indicates how well the independent variables in the model (the causes) explain the dependent variable. The adjusted R^2 is actually a percentage of variance explained by the independent variables. A good, nontechnical way to think about it is that R^2 tells us what percent of the variation in the dependent variable is explained. In the education and income example, $R^2 =$ 0.76, which means that education explains 76 percent of income.

Finally, the regression method allows researchers to sort out the effects of two or more independent causes simultaneously. Instead of just looking at the influence of education on income, we can look at the influence of education, race, gender, and other variables—all at the same time. The key here is that regression yields estimates of the effect of each independent variable while holding the effects of all the others constant. In the regression equation below, we see a hypothetical example of the effects of education and age on income:

$$\text{Income} = \$2{,}500 + \$2{,}000 \ (\text{Education}) + \$600 \ (\text{Age})$$

In this example, as people get older, they get paid more. So our hypothetical high school graduate would be expected to earn about six hundred dollars more each year because of seniority, growing experience, or other factors.

FACTOR ANALYSIS

"Factor analysis" is a method used to search for and measure unobserved variables, called *factors*. For example, consider the problem of measuring the mathematical ability of students. Some people are good at math; others are not. We believe that mathematical ability exists, but we cannot directly observe or measure it. Factor analysis offers a method for doing this.

Figure A2 illustrates how the model works. In the case of math ability, we can ask students to answer a number of math questions. The unobserved factor (in the circle) is assumed to cause the observed answers to the test questions (the boxed "questions"). The factor-analysis method makes use of correlations among observed variables to calculate a set of causal coefficients, or *factor loadings*, similar to standardized regression coefficients. The loadings generally range from -1 to $+1$. Zero indicates no association. Positive loadings indicate that as the value of the factor (say, math ability) increases, the student is more likely to get the test question right. Negative loadings indicate the opposite. If we were to conduct a factor analysis on a set of questions in a math test, we should expect to get mostly large, positive factor loadings (perhaps .80 or so).

Factor analysis can actually estimate more than one factor at a time. For example, instead of just including math questions on a test, suppose we included some English grammar and syntax questions as well. If language ability were distinct from math ability, the factor analysis method would report two factors and a set of factor loadings for both factors. In this case, the math questions would have high loadings for the math factor, but lower loadings for the lan-

Figure A2 A Factor Analysis

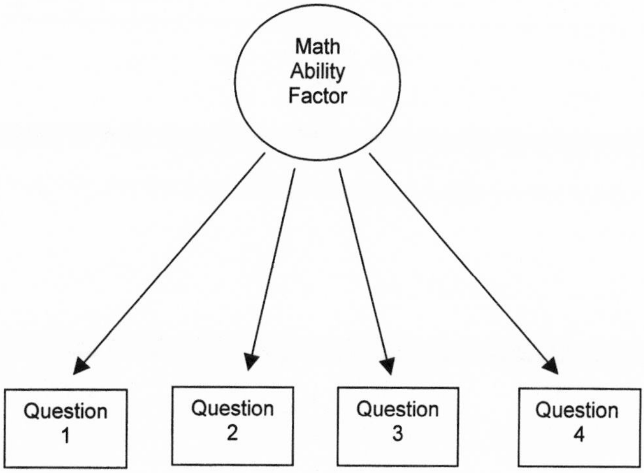

guage-ability factor. If language and math ability were completely unrelated, the math questions would have high loadings on the math factor and loadings close to zero for the language factor.

We can use this method to test questions being considered for use in building a scale. If we have a set of six information questions, and five of them have high loadings on a factor while the sixth has a very small loading, we can conclude that the first five items would make a good test and that we should drop the sixth question.

Data Appendix

A variety of data sets are used in this book. Although all data sets are identified in the endnotes, this appendix provides additional information about them.

In chapter 2, the data in figures 2.2 through 2.8 are from the U.S. Energy Department, Energy Information Administration's *Annual Energy Review 1999*. This report, which offers a rich variety of data on our energy situation, is available on the World Wide Web at www.eia.doe.gov/aer/.

In chapter 3, the U.S. data in figure 3.1 and 3.3, and tables 3.3–3.5 are from surveys conducted by Cambridge Reports/Research International (955 Massachusetts Ave., Cambridge, Massachusetts, 02139). The minimum sample size is nine hundred. The 1974–1988 data are from William G. Mayer, *The Changing American Mind: How and Why American Public Opinion Changed between 1960 and 1988* (Ann Arbor: University of Michigan Press, 1992), 350, 355, 489, 492. The 1989–1990 data on nuclear power are from Eugene A. Rosa and Riley E. Dunlap, "Nuclear Power: Three Decades of Public Opinion," *Public Opinion Quarterly* 58 (Summer 1994): 295–325.

The California data in figure 3.2 are from surveys conducted by the Field Institute, a nonpartisan, nonprofit public opinion research organization established by the Field Research Corporation (550 Kearny Street, Suite 900, San Francisco, California, 94108). The Field Polls were written by Field Institute staff; the 1998 California Offshore Oil Drilling and Energy Policy Survey was written by the author. The samples were representative cross-sections of California adults. The Field Institute labels its polls by year and survey number, so "7703" refers to the third survey conducted in 1977. The Field Polls used in figure 3.2, their dates, and sample sizes are shown in the table.

The California data in tables 3.1 and 3.2 are from Field Polls 8104 and 9004, and the California Offshore Oil Drilling and Energy Policy Survey.

In chapter 4, the data in tables 4.2, 4.3, and 4.6, and figures 4.2 and 4.3 are from the California Offshore Oil Drilling and Energy Policy Survey.

Field Polls Dates and Sample Sizes

Survey	Dates	Sample Size
7703	June 17–July 2, 1977	1,034
7801	January 7–15, 1978	1,003
7902	May 3–15, 1979	485 (random half; total sample N = 979)
8002	April 2–8, 1980	501 (random half; total sample N = 1,012)
8006	October 15–18, 1980	506 (random half; total sample N = 1,018)
8104	October 26–November 1, 1981	1,102
8401	February 1–9, 1984	743 (random half; total sample N = 1,511)
8903	July 12–23, 1989	993
9004	August 17–27, 1990	614 (random half; total sample N = 1,235)
COODEPS*	March 5–18, 1998	810

*California Offshore Oil Drilling and Energy Policy Survey

The 1977 questions used for tables 4.5 and 4.7 were:

Now I'm going to read you a series of statements about the energy situation and ways that have been suggested to take care of it. I'd like you to tell whether you agree or disagree with each of the statements as I read it. Here's the first one . . .

A. People who buy cars that get low gas mileage should pay an extra tax, while people who buy cars that get very high gas mileage should get a tax rebate.

B. Homeowners who insulate their homes should get a tax credit to pay part of the cost.

C. People who install solar energy devices in their homes should get tax credits to help pay part of the cost.

D. Gasoline taxes should be increased to get people to drive their cars less than they do now.

E. More tax revenues should be used to help subsidize the cost of improved public mass transportation facilities.

F. It would be better to cut back on living standards in order to conserve energy rather than to go on using up natural resources at the present rate.

G. Solar energy could solve the energy shortage within the next few years if it was given higher priority for research and development.

H. Oil and gas prices should be controlled, but tax breaks should be allowed to encourage more oil and gas exploration and drilling in the U.S.

I. The price of oil and gas produced in the U.S. should be decontrolled and allowed to go as high as necessary to encourage more exploration and drilling for oil and gas in the U.S.

J. Air pollution standards should be reduced somewhat to permit the substitution of coal, which is more plentiful in this country than oil and gas.

K. California should allow oil companies to drill more oil wells in state tidelands along the seacoast.

L. California should allow oil companies to build more tanker terminals for unloading oil and gas tankers in California.

M. California should allow oil companies to build a terminal near Los Angeles or San Francisco for unloading tankers carrying Liquefied Natural Gas.
N. More nuclear power plants should be constructed to meet future energy needs.

The 1998 questions used for tables 4.6 and 4.8 were:

I would like to start by reading you a series of statements about the energy situation. I'd like you to tell me whether you agree strongly, agree slightly, disagree slightly, or disagree strongly with each of the statements as I read it. Here's the first one . . .
A. Energy taxes should be increased to get businesses to use energy more efficiently.
B. People who buy cars that get low gas mileage should pay an extra tax, while people who buy cars that get very high gas mileage should get a tax rebate.
C. I would prefer to cut back on my standard of living in order to conserve energy rather than to go on using up natural resources at the present rate.
D. Population growth and housing development in California should be slowed down to reduce energy needs.
E. The growth of industries requiring large amounts of energy should be slowed down to reduce energy needs.
F. Oil companies should be allowed to drill more oil and gas wells in state tidelands along the California seacoast.
G. Current government restrictions prohibiting the drilling of oil and gas wells on government parklands and forest reserves should be relaxed.
H. The building of more nuclear power plants should be allowed in California.

Chapters 4 and 5 use a slightly modified version of Delli Carpini and Keeter's political knowledge index, shown below. Because the survey was conducted in the spring, rather than immediately after the November election, question D asks about the current majority in the House instead of the majority before the election. For question E, a randomly selected half of the sample was asked which party is more conservative; the other half was asked which party is more liberal.

Last, here are a few questions about the government in Washington. Many people don't know the answers to these questions, so if there are some you don't know, just tell me and we'll go on.
A. Do you know what job or political office is now held by Al Gore?
B. Whose responsibility is it to determine if a law is constitutional or not . . . is it the president, the Congress, or the Supreme Court?
C. How much of a majority is required for the U.S. Senate and House to override a presidential veto?
D. Do you happen to know which party has the most members in the House of Representatives right now?
E. Would you say that one of the parties is more [conservative/liberal] than the other at the national level? Which party is more [conservative/liberal?}

The three surveys that are the focus of chapter 5 are Field Polls 8104 and 9004, and the California Offshore Oil Drilling and Energy Policy Survey, which are described above.

References

Abramson, Paul R. *Political Attitudes in America: Formation and Change.* San Francisco: Freeman, 1983.

Abramson, Paul R., and Ronald Inglehart. *Value Change in Global Perspective.* Ann Arbor: University of Michigan Press, 1995.

Achen, Christopher H. "Mass Political Attitudes and the Survey Response." *American Political Science Review* 69 (1975): 1218–31.

———. "Measuring Representation." *American Journal of Political Science* 22 (1978): 475–510.

———. "What Does 'Explained Variance' Explain? Reply." *Political Analysis* 2 (1990): 173–84.

Adams, Glenn. "Federal-Ordered Destruction of Dam May Set a Precedent." *Santa Barbara News-Press,* 2 July 1999, A8.

Ahern, William R. "California Meets the LNG Terminal." *Coastal Zone Management Journal* 7 (1980): 185–221.

Arcury, Thomas A. "Environmental Attitude and Environmental Knowledge." *Human Organization* 49 (Winter 1990): 300–304.

Arcury, Thomas A., and T. P. Johnson. "Public Environmental Knowledge: A Statewide Survey." *Journal of Environmental Education* 18: 31–37.

Arcury, Thomas A., T. P. Johnson, and S. J. Scollay. "Ecological Worldview and Environmental Knowledge: An Examination of the New Environmental Paradigm." *Journal of Environmental Education* 17: 35–40.

———. "Sex Differences in Environmental Concern and Knowledge: The Case of Acid Rain." *Sex Roles* 16: 463–72.

Arnold, R. Douglas. *The Logic of Congressional Action.* New Haven, Conn.: Yale University Press, 1990.

Arrhenius, Svante. "On the Influence of Carbolic Acid in the Air upon the Temperature of the Ground." *Philosophical Magazine* 41 (April 1896): 237–77.

Baird, Brian N. R. "Tolerance for Environmental Health Risks: The Influence of Knowledge, Benefits, Voluntariness, and Environmental Attitudes." *Risk Analysis* 6 (1986): 425–35.

Barabak, Mark Z. "Most Californians Think Electricity Crunch Is Artificial." *Los Angeles Times,* 7 January 2001, A1.

Bartley, Ernest B. *The Tidelands Oil Controversy: A Legal and Historical Analysis* Austin: University of Texas Press, 1953.

Baxter, Sandra, and Marjorie Lansing. *Women and Politics: The Visible Majority.* Ann Arbor: University of Michigan Press, 1986.

Beckel, Robert G. "Presidential Politics Infuses Gas-Tax Debate." *Los Angeles Times,* 12 May 1996, M2.

Bennett, Stephen E. "Attitude Structures and Foreign Policy Options." *Social Science Quarterly* 55 (1974): 732–42.

———. "Consistency among the Public's Social Welfare Policy Attitudes in the 1960's." *American Journal of Political Science* 17 (1973): 544–70.

———. "Trends in Americans' Political Information, 1967–1987." *American Politics Quarterly* 17 (1989): 422–35.

Benson, Edward G. "Three Words." *Public Opinion Quarterly* 4 (1940): 130–34.

Berry, John M., and Eric Pianin. "Panel Finds Price Index as Excessive." *Washington Post,* 5 December 1996.

Berry, Mary Clay. *The Alaska Pipeline: The Politics of Oil and Native Land Claims.* Bloomington: Indiana University Press, 1975.

"Beyond 'The China Syndrome.' " *Newsweek,* 16 April 1979, 31.

Bishop, George D., David L. Hamilton, and John B. McConahay. "Attitudes and Nonattitudes in the Belief Systems of the Mass Public." *Journal of Social Psychology* 110 (1980): 53–64.

Bishop, George F. "The Effect of Education on Ideological Consistency." *Public Opinion Quarterly* 40 (1976): 337–48.

Bishop, George F., Robert W. Oldendick, and Alfred J. Tuchfarber. "Change in the Structure of American Political Attitudes: The Nagging Question of Question Wording." *American Journal of Political Science* 22 (1978): 250–69.

Bloom, Saul, John M. Miller, James Warner, and Philippa Winkler. *Hidden Casualties: Environmental, Health, and Political Consequences of the Persian Gulf War.* Berkeley, Calif.: North Atlantic, 1994.

Bohle, Robert H. "Negativism as News Selection Predictor." *Journalism Quarterly* 63 (1986): 789–96.

Bord, Richard J., Robert E. O'Connor, and Ann Fisher. "In What Sense Does the Public Need to Understand Global Warming?" *Public Understanding of Science* 9 (2000): 205–18.

Bostrom, Ann, et al. "What Do People Know about Global Climate Change?" *Risk Analysis* 14 (1994): 959–69.

Brooks, Nancy Rivera, and Jacqueline Newmyer. "Electricity Deregulation a Shock to the System: Regulators Act to Slow Rate Hikes but Reject Freeze." *Los Angeles Times,* 4 August 20, 2000, A1.

Brownstein, Ronald. "Democrats Reassert Role of Government in Marketplace." *Los Angeles Times,* 2 May 1996.

Bruni, Frank. "Bush, in Energy Plan, Endorses New U.S. Drilling to Remedy Oil Prices." *New York Times,* 30 September 2000.

Burger, Joanna. *Oil Spills.* New Brunswick, N.J.: Rutgers University Press, 1997.

Buttel, Frederick H. "Age and Environmental Concern: A Multivariate Analysis." *Youth and Society* 10 (March 1979): 237–56.

Buttel, Frederick H., and William L. Flinn. "The Structure of Support for the Environmental Movement, 1968–1970." *Rural Sociology* 39 (1974): 56–69.

Caldwell, Andy. "Dealing with Facts, Figures Involving Local Oil Industry." *Santa Barbara News-Press,* 25 February 1996.

Campbell, Colin J. *The Coming Oil Crisis.* Essex: Multi-Science, 1998.

———. *The Golden Century of Oil, 1950–2050.* Dordrecht, Neth.: Kluwer Academic, 1991.

Campbell, Colin J., and Jean H. Laherrère. "The End of Cheap Oil." *Scientific American,* March 1998, 78–83.

Carmines, Edward G., and James A. Stimson. *Issue Evolution: Race and the Transformation of American Politics.* Princeton, N.J.: Princeton University Press, 1989.

Carmines, Edward G., and Richard A. Zeller. *Reliability and Validity Assessment.* Beverly Hills, Calif.: Sage, 1979.

Carson, Rachel. *Silent Spring.* Boston: Houghton Mifflin, 1962.

"Cheney Sees Benefits in Nuclear Plants." *Los Angeles Times,* 22 March 2001, A3.

Cicchetti, Charles J. *Alaskan Oil: Alternative Routes and Markets* Washington, D.C.: Resources for the Future, 1972.

Cicin-Sain, Biliana. "California and Ocean Management: Problems and Opportunities." *Coastal Management* 18 (1990): 311–35.

Cline, William R. *The Economics of Global Warming.* Washington, D.C.: Institute for International Economics, 1992.

Cohen, Michael P. *The History of the Sierra Club, 1892–1970.* San Francisco: Sierra Club, 1988.

Cone, Marla, and Gary Polakovic. "Bush's Idea of Easing Smog Rules Won't Help, Experts Say." *Los Angeles Times,* 25 January 2001, A18.

Congressional Quarterly. *Congressional Quarterly Almanac, 102nd Congress.* Washington, D.C.: Congressional Quarterly, 1993.

———. *Energy and Environment: The Unfinished Business.* Washington, D.C.: Congressional Quarterly, 1985.

Converse, Philip E. "The Nature of Belief Systems in Mass Publics." In *Ideology and Discontent,* edited by David Apter. New York: Free Press, 1964.

———. "Attitudes and Nonattitudes: The Continuation of a Dialogue." In *The Quantitative Analysis of Social Problems,* edited by Edward Tufte. Reading, Mass.: Addison-Wesley, 1970.

Cook, Elizabeth Adell, Ted G. Jelen, and Clyde Wilcox. *Between Two Absolutes.* Boulder, Colo.: Westview, 1992.

Couch, Arthur, and Kenneth Keniston. "Yeasayers and Naysayers: Agreeing Response Set as a Personality Variable." *Journal of Abnormal and Social Psychology* 60 (1960): 151–74.

"Crude Oozes Its Way Back to $30 a Barrel; Supply Boost Unlikely." *Los Angeles Times,* 13 May 2000, C1.

Cunningham, W. H., and B. Joseph. "Energy Conservation, Price Increases, and Payback Periods." In *Advances in Consumer Research,* vol. 5, edited by H. H. Keith. Ann Arbor, Mich.: Association for Consumer Research, 1978, 201–205.

Cutter, Susan Caris, Community Concern for Pollution: Social and Environmental Influences." *Environment and Behavior* 13 (January 1981): 105–24.

Davidson, Debra J., and William R. Freudenburg. "Gender and Environmental Concerns: A Review and Analysis of Available Research." *Environment and Behavior* 28 (1996): 302–39.

Davis, David H. *Energy Politics.* 2d ed. New York: St. Martin's, 1978.

Davis, James A. "Communism, Conformity, Cohorts, and Categories: American Tolerance in 1954 and 1972–73." *American Journal of Sociology* 81 (1975): 491–513.

————. "Review Essay on Paul R. Abramson and Ronald Inglehart, *Value Change in Global Perspective.*" *Public Opinion Quarterly* 60 (1996): 322–31.

Dean, Gillian, and Thomas W. Moran. "Measuring Mass Political Attitudes: Change and Unreliability." *Political Methodology* 4 (1977): 383–413.

Delli Carpini, Michael X., and Scott Keeter. "Stability and Change in the U.S. Public's Knowledge of Politics." *Public Opinion Quarterly* 55 (1991): 583–612.

————. *What Americans Know about Politics and Why It Matters.* New Haven, Conn.: Yale University Press, 1996.

Devall, William B. "Conservation: An Upper-Middle Class Social Movement: A Replication." *Journal of Leisure Research* 2 (1970): 123–25.

Diamond, Edwin. *The Tin Kazoo.* Cambridge, Mass.: MIT Press, 1975.

Dickerson, Marla. "Manufacturers Reeling from Soaring Price of Natural Gas." *Los Angeles Times,* 17 December 2000, A1.

Donn, Jeff. "Nuclear Energy Running Out of Steam." *Santa Barbara News-Press,* 28 March 1999, A7.

Douglas, Mary. *Risk and Blame: Essays in Cultural Theory.* London: Routledge, 1992.

Douglas, Mary, and Aaron Wildavsky. *Risk and Culture.* Berkeley: University of California Press, 1982.

Dowling, Edward T., and Francis Gittilton. "Oil in the 1980s: An OECD Perspective." In *The Oil Market in the 1980s: A Decade of Decline,* eds. Siamack Shojai and Bernard S. Katz. New York: Praeger, 1992.

Dunlap, Riley E. "The Evolution of the U.S. Environmental Movement from 1970 to 1990: An Overview." In *American Environmentalism: The U.S. Environmental Movement, 1970–1990,* edited by Riley E. Dunlap and Angela G. Mertig. Philadelphia: Taylor and Francis, 1992, 1–10.

————. "Public Opinion and Environmental Policy." In *Environmental Politics and Policy: Theory and Evidence,* edited by James P. Lester. 2d ed. Durham, N.C.: Duke University Press, 1995.

————. "Trends in Public Opinion toward Environmental Issues." In *American Environmentalism: The U.S. Environmental Movement, 1970–1990,* edited by Riley E. Dunlap and Angela G. Mertig. Philadelphia: Taylor and Francis, 1992, 89–116.

Dunlap, Riley E., and Kent D. Van Liere. "The New Environmental Paradigm: A Proposed Measuring Instrument and Preliminary Results." *Journal of Environmental Education* 9 (1978): 10–19.

Dunlap, Riley E., and Rik Scarce. "Trends: Environmental Problems and Protections." *Public Opinion Quarterly* 55 (1991): 651–72.

Egelko, Bob. "Conservative Justice's Opinions Raise Eyebrows." *Santa Barbara News-Press,* 19 October 1999, B4.

Eiser, J. Richard, Russel Spears, Paul Webey, and Joop van der Pligt. "Local Residents' Attitudes to Oil and Nuclear Developments." *Social Behavior* 3 (1988): 237–53.

Ellis, Richard J., and Fred Thompson. "Culture and the Environment in the Pacific Northwest." *American Political Science Review* 91 (December 1997): 885–97.

Energy Information Agency. *Annual Energy Review 1997.* Washington, D.C.: Government Printing Office, 1997.

"Energy: No Shortage of Suspicions." *Newsweek,* 14 January 1974, 63–64.

"Enron's Profit Surge Powers Energy Stocks." *Los Angeles Times,* 23 January 2001, C1.

Erfle, Stephen, and Henry McMillan. "Determinants of Network News Coverage of the Oil Industry during the Late 1970s." *Journalism Quarterly* 66 (1989): 121–28.

Esposito, John C., et al. *Vanishing Air.* New York: Pantheon, 1970.

"Facing Up to Cold Reality." *Newsweek,* 19 November 1973, 109–110.

Farhar-Pilgrim, Barbara, and William R. Freudenburg. "Nuclear Energy in Perspective: A Comparative Assessment of the Public View." In *Public Reaction to Nuclear Power: Are There Critical Masses?* edited by William R. Freudenburg and Eugene A. Rosa. Boulder, Colo.: Westview, 1984, 183–203.

Feldman, Stanley. "Measuring Issue Preferences: The Problem of Response Instability." *Political Analysis* 1 (1989): 25–60.

———. "Structure and Consistency in Public Opinion: The Role of Core Beliefs and Values." *American Journal of Political Science* 32 (May 1988): 416–40.

Feldman, Stanley, and John Zaller. "The Political Culture of Ambivalence: Ideological Responses to the Welfare State." *American Journal of Political Science* 36 (February 1992): 268–307.

Fischhoff, Baruch, Paul Slovic, and Sarah Lichtenstein. "Poorly Thought-Out Values: Problems of Measurement." In *Energy and Material Resources: Attitudes, Values, and Public Policy,* edited by W. David Conn. American Association for the Advancement of Science Selected Symposia series, no. 75. Boulder, Colo.: Westview, 1983.

Fort, Rodney. "The Decline of Nuclear Power in the United States: Inherent versus Economic Anti-Nuclear Sentiment." In *Nuclear Power at the Crossroads,* edited by Thomas C. Lowinger and George W. Hinman. Boulder, Colo.: International Research Center for Energy, 1994, 165–89.

Freudenburg, William R. "Rural-Urban Differences in Environmental Concern: A Closer Look." *Rural Sociology* 61 (May 1991): 167–98.

Freudenburg, William R., and Eugene A. Rosa, eds. *Public Reactions to Nuclear Power: Are There Critical Masses?* Boulder, Colo.: Westview, 1984.

Freudenburg, William R., and Robert Gramling. *Oil in Troubled Waters: Perceptions, Politics, and the Battle over Offshore Oil Drilling.* Albany: State University of New York Press, 1994.

Fulwood, Sam, III. "Senate Strikes Deal on Minimum Wage, Gas Tax." *Los Angeles Times,* 26 June 1996, A1.

Gallup, George H. *The Gallup Poll: Public Opinion 1935–1971.* New York: Random House, 1972.

George, David L., and Priscilla L. Southwell. "Opinion on Diablo Canyon Nuclear Power Plant: The Effects of Situation and Socialization." *Social Science Quarterly* 67 (1986): 722–35.

George, Philip. *The Political Economy of International Oil.* Edinburgh: Edinburgh Press, 1994.

Ginsberg, Benjamin, and Alan Stone, eds., *Do Elections Matter?* Armonk, N.Y.: Sharpe, 1996.

Golob, Richard, and Eric Brus. *The Almanac of Renewable Energy: The Complete Guide to Emerging Energy Technology.* New York: Henry Holt, 1993.

Gosselin, Peter G. "Most of the West in the Same Power Jam as California." *Los Angeles Times,* 26 February 2001, A1.

Gramling, Robert. *Oil on the Edge.* Albany: State University of New York Press, 1996.

Hagner, Paul R. and John C. Pierce. "Levels of Conceptualization and Political Belief Consistency." *Micropolitics* 2 (1983): 311–48.

Harris, Louis. "A Call for Tougher—Not Weaker—Antipollution Laws." *Business Week*, 24 January 1983, 87.

———. "Energy Crisis Has Many Causes." *Chicago Tribune*, 11 August 1977.

———. "Energy Shortages Regarded as Serious by Most Americans." *Chicago Tribune*, 26 July 1973.

Harry, Joseph, Richard Gale, and John Hendee. "Conservation: An Upper-Middle Class Social Movement." *Journal of Leisure Research* 1 (1969): 246–54.

Hawley, T. M. *Against the Fires of Hell: The Environmental Disaster of the Gulf War*. New York: Harcourt, Brace, Jovanovich, 1992.

Hempel, Lamont C. *Environmental Governance: The Global Challenge*. Washington, D.C.: Island, 1996.

Hensler, Deborah R., and Carl P. Hensler. "Evaluating Nuclear Power: Voter Choice on the California Nuclear Energy Initiative." Santa Monica, Calif.: Rand, 1979.

Hershey, Marjorie Randon, and David B. Hill. "Is Pollution 'A White Thing'? Racial Differences in Preadults' Attitudes." *Public Opinion Quarterly* 41 (1977–78): 439–58.

Hirsh, Michael. "Getting All Pumped Up." *Newsweek*, 13 May 1996, 48.

Hogan, William W. "Patterns of Energy Use." In *Energy Conservation: Successes and Failures*, eds. John C. Sawhill and Richard Colton. Washington, D.C.: Brookings Institution, 1986.

Honnold, Julie A. "Age and Environmental Concern." *Journal of Environmental Education* 16, no. 1 (Fall 1984): 4–9.

Hook, Janet. "House Votes to Repeat 4.3-Cent Gas Tax Hike." *Los Angeles Times*, 22 May 1996, A1.

Horwitch, Mel. "Coal: Constrained Abundance." In *Energy Future: Report of the Energy Project at the Harvard Business School*, eds. Robert Stobaugh and Daniel Yergin. New York: Random House, 1979, chap. 4.

Houghton, John. *Global Warming: The Complete Briefing*. 2d ed. Cambridge: Cambridge University Press, 1997.

Houts, Peter S., Paul D. Cleary, and Teh-Wei Hu. *The Three Mile Island Crisis: Psychological, Social, and Economic Impacts on the Surrounding Population*. University Park: Pennsylvania State University Press, 1988.

Inglehart, Ronald. *The Silent Revolution: Changing Values and Political Styles among Western Publics*. Princeton, N.J.: Princeton University Press, 1977.

———. "Value Change in Industrial Societies." *American Political Science Review* 81 (December 1987): 1289–1303.

———. "Value Priorities and Social Change." In *Political Action: Mass Participation in Five Western Democracies*, edited by Samuel H. Barnes and Max Kaase. Beverly Hills, Calif.: Sage, 1979, 305–42.

———. *Culture Shift in Advanced Industrial Society*. Princeton, N.J.: Princeton University Press, 1990.

Ingram, Carl, and Nancy Vogel. "Legislature OKs San Diego Electric Relief Package." *Los Angeles Times*, 31 August 2000, A1.

"Intermission: 'Too Late and Too Early.'" *Newsweek*, 12 June 1967, 38–40.

Isser, Steve. *The Economics and Politics of the United States Oil Industry, 1920–1990*. New York: Garland, 1996.

Jackman, Robert W. "Prejudice, Tolerance, and Attitudes toward Ethnic Groups." *Journal of Politics* 34 (1972): 753–73.

Jackson, Thomas H., and George E. Marcus. "Political Competence and Ideological Constraint." *Social Science Research* 4 (1975): 93–111.

Jennings, M. Kent. "Ideology among Mass Publics and Political Elites." *Public Opinion Quarterly* 56 (1992): 420–41.

Jennings, M. Kent, and Richard G. Niemi. *Generations and Politics: A Panel Study of Young Adults and Their Parents.* Princeton, N.J.: Princeton University Press, 1981.

———. *The Political Character of Adolescence.* Princeton, N.J.: Princeton University Press, 1974.

Jones, Charles O. *Clean Air: The Policies and Politics of Pollution Control.* Pittsburgh, Pa.: University of Pittsburgh Press, 1975.

Jones, G. Kevin. "The Development of Outer Continental Shelf Oil and Gas Resources." In *Energy Resources Development: Politics and Policies,* edited by Richard L. Ender and John Choon Kim. New York: Quorum, 1987.

Jones, Robert Emmet, and Lewis F. Carter. "Concern for the Environment among Black Americans: An Assessment of Common Assumptions." *Social Science Quarterly* 75 (September 1994): 560–79.

Jones, Robert Emmet, and Riley E. Dunlap. "The Social Bases of Environmental Concern: Have They Changed over Time?" *Rural Sociology* 57 (1992): 28–47.

Kalt, Joseph P., and Mark A. Zupan. "Further Evidence on Capture and Ideology in the Economic Theory of Politics. *American Economic Review* 74 (1984): 279–300.

Kamieniecki, Sheldon. "Political Parties and Environmental Policy." In *Environmental Politics and Policy,* edited by James P. Lester. 2d ed. Durham, N.C.: Duke University Press, 1995.

Katz, James Everett. *Congress and National Energy Policy.* New Brunswick, N.J.: Transaction, 1984.

Keith, Bruce E., David B. Magleby, Candice J. Nelson, Elizabeth Orr, Mark Westlye, and Raymond E. Wolfinger. *The Myth of the Independent Voter.* Berkeley: University of California Press, 1992.

Keith, Leon D. "Pressure Building for More Dams." *Santa Barbara News-Press,* 15 April 2000, 4.

Kempton, Willett. "Public Understanding of Global Warming." *Society and Natural Resources* 4 (1991): 331–45.

Kempton, Willet, James S. Boster, and Jennifer A. Hartley, *Environmental Values and American Culture.* Cambridge, Mass.: MIT Press, 1995.

Kennedy, Harold W., and Martin E. Weekes. "Control of Automobile Emissions: California Experience and the Federal Legislation." In *Air Pollution Control,* edited by Clark C. Havighurst. Dobbs Ferry, N.Y.: Oceana, 1969, 101–18.

Kerr, Jennifer. "Voters Would Block Power Rate Hike, Davis Warns Wall Street." *Santa Barbara News-Press,* 3 March 2001, A3.

King, Gary. "Stochastic Variation: A Comment on Lewis-Beck and Skalaban's 'The R-Squared.' " *Political Analysis* 2 (1990): 185–200.

Kingdon, John. *Congressmen's Voting Decisions.* New York: Harper and Row, 1981.

Kowalewski, David. "Environmental Attitudes in Town and Country: A Community Survey." *Environmental Politics* 3 (Summer 1994): 295–311.

Kraul, Chris. "Antiquated Power Lines Add to Energy Woes." *Los Angeles Times,* 31 January 2001, A1.

————. "Charges of Gouging as Power Costs Skyrocket." *Los Angeles Times*, 8 August 2000, A1.

————. "Natural Gas Prices Rise as Worries over Energy Build." *Los Angeles Times*, 22 August 2000, A1.

————. "Oil Prices Jump Amid Renewed Mideast Troubles." *Los Angeles Times*, 13 October 2000, C1.

————. "Power Crisis Generates Windfall for Suppliers." *Los Angeles Times*, 27 December 2000, A1.

Krosnick, Jon A. "Government Policy and Citizen Passion: A Study of Issue Publics in Contemporary America." *Political Behavior* 12 (March 1990): 59–92.

Kuklinski, James H., Daniel S. Metlay, and W. D. Kay. "Citizen Knowledge and Choices on the Complex Issue of Nuclear Energy." *American Journal of Political Science* 26 (1982): 615–42.

Landy, Marc K., J. Roberts, and Stephen R. Thomas. *The Environmental Protection Agency.* Expanded ed. New York: Oxford University Press, 1994.

Lankford, John. "Vote Split on Goleta as City, Polls Show." *Santa Barbara News-Press*, 23 October 1990, A1.

Lee, Patrick. "Anger Flares as Oil Officials Defend Gasoline Price Hikes." *Los Angeles Times*, 26 April 1996, A1.

————. "Texaco Led Run-up in Southland Gas Prices." *Los Angeles Times,* 7 May 1996, A1.

Lee, Thomas H., Ben C. Ball, Jr., and Richard D. Tabors. *Energy Aftermath.* Boston: Harvard Business School Press, 1990.

Leonhardt, David, and Barbara Whitaker. "Higher Fuel Prices Do Little to Alter Motorists' Habits." *New York Times,* 10 October 2000.

LePage, Andrew. "Both Sides Spouting Off on Oil Issue." *Santa Barbara News-Press,* 2 March 1996, A1.

LePage, Andrew, and Melinda Burns, "Mobil Files Clearview Application." *Santa Barbara News-Press,* 10 February 1995, A1.

Lewis-Beck, Michael S., and Andrew Skalaban. "The R-Squared: Some Straight Talk." *Political Analysis* 2 (1990): 153–72.

Lipset, Seymour Martin, and William Schneider. *The Confidence Gap: Business, Labor, and Government in the Public Mind.* Baltimore: Johns Hopkins University Press, 1987.

Lord, Frederick M., and Melvin R. Novick. *Statistical Theories of Mental Test Scores.* Reading, Mass.: Addison Wesley, 1968.

Lovins, Amory B. *Soft Energy Paths.* Cambridge, Mass.: Ballinger, 1977.

————. *World Energy Strategies.* Cambridge, Mass.: Ballinger, 1971.

Lovins, Amory B., and L. Hunter Lovins. *Energy War: Breaking the Nuclear Link.* New York: Harper and Row, 1980.

Luttbeg, Norman R., and Michael M. Gant. "The Failure of Liberal/Conservative Ideology as a Cognitive Structure." *Public Opinion Quarterly* 49 (1985): 80–93.

Lyman, Rick. "Power Shortage Fueling Cynicism." *Santa Barbara News-Press*, 13 January 2001, A1.

Maddala, G. S. *Introduction to Econometrics.* New York: Macmillan, 1988.

Maharik, M., and Baruch Fischhoff. "Contrasting Perceptions of the Risks of Using Nuclear Energy Sources in Space." *Journal of Environmental Psychology* 13 (1993): 243–50.

Marris, Claire, Ian H. Langford, and Timothy O'Riordan. "A Quantitative Test of the Cultural Theory of Risk Perceptions: Comparison with the Psychometric Paradigm." *Risk Analysis* 18 (1998): 635–47.

Marshall, Tyler. "U.S. Dives into a Sea of Major Rewards—and Risks." *Los Angeles Times*, 23 February 1998.

Maslow, Abraham. *Motivation and Personality*. New York: Harper and Row, 1954.

Mayer, William G. *The Changing American Mind: How and Why American Public Opinion Changed between 1960 and 1988*. Ann Arbor: University of Michigan Press, 1992.

McClosky, Herbert, and John Zaller. *The American Ethos: Public Attitudes toward Capitalism and Democracy*. Cambridge, Mass.: Harvard University Press, 1984, chap. 8 (coauthor Dennis Chong).

McEvoy, James. "The American Concern with the Environment." In *Social Behavior, Natural Resources, and the Environment*, edited by W. R. Burch et al. New York: Harper and Row, 1972.

Meagher, Ed. "Liquified Natural Gas—Risk of a Disaster Feared." *Los Angeles Times*, 12 January 1975.

Merkleim, Helmuut A., and W. Carey Hardy. *Energy Economics*. Houston: Gulf, 1977.

Mezey, Susan Gluck. "Attitudinal Consistency among Political Elites: Implications of Support for the Equal Rights Amendment." *American Politics Quarterly* 9 (1981): 111–25.

"Middle East: The Scent of War." *Newsweek*, 5 June 1967, 40–48.

Miller, Arthur. "Gender and the Vote: 1984." In *The Politics of the Gender Gap*, edited by Carol M. Mueller. Newbury Park, Calif.: Sage, 1988.

Miller, Warren E., and Donald E. Stokes. "Constituency Influence in Congress." In *Elections and the Political Order*, edited by Angus Campbell, Philip E. Converse, Warren E. Miller, and Donald E. Stokes. New York: Wiley, 1966.

Mitchell, Alison. "Gore Says Bush Plan Will Cause Lasting Damage to the Environment." *New York Times*, 30 September 2000.

Mitchell, Robert C. "Public Responses to a Major Failure of a Controversial Technology." In *Accident at Three Mile Island: The Human Dimensions*, edited by David L. Sills et al. Boulder, Colo.: Westview, 1982, 21–38.

Mohai, Paul. "Black Environmentalism." *Social Science Quarterly* 71 (December 1990): 744–65.

Mohai, Paul, and Ben W. Twight. "Age and Environmentalism: An Elaboration of the Buttel Model Using National Survey Evidence." *Social Science Quarterly* 68 (1987): 798–815.

Molotch, Harvey. "Oil in Santa Barbara and Power in America." *Sociological Inquiry* 40 (Winter 1970): 131–44.

Molotch, Harvey, and William Freudenburg, eds. *Santa Barbara County: Two Paths*. Final Report to the Minerals Management Service [MMS], OCS Study MMS 96-0036. Camarillo, Calif.: U.S. Minerals Management Service, 1996.

Morain, Dan. "Davis Walks a Political Tightrope on Energy Prices." *Los Angeles Times*, 12 September 2000, A1.

Morone, Joseph G., and Edward J. Woodhouse. *Averting Catastrophe: Strategies for Regulating Risky Technologies*. Berkeley: University of California Press, 1986.

———. *The Demise of Nuclear Energy? Lessons for Democratic Control of Technology*. New Haven, Conn.: Yale University Press, 1989.

Morrison, D. E., K. E. Hornback, and W. K. Warner. "The Environmental Movement: Some Preliminary Observations and Predictions." In *Social Behavior, Natural Resources*, *and the Environment*, edited by W. R. Burch, Jr., et al. New York: Harper and Row, 1972, 259–79.

Morrison, Denton E., and Riley E. Dunlap. "Environmentalism and Elitism: A Conceptual and Empirical Analysis." *Environmental Management* 10 (1986): 581–89.

Mulligan, Thomas S. "Tech Companies a Drain on Power Grid." *Los Angeles Times*, 12 December 2000, A1.

Murkowski, Frank H. "Let Alaskan Oil Help the State, Nation." *Los Angeles Times*, 17 February 2000, B9.

Murphy, Kim. "Between a Rock and a Hard Place: Where Dams, Salmon Meet." *Los Angeles Times*, 13 June 2000, A5.

———. "Talk of Demolishing Dams Yields Torrents of Debate." *Los Angeles Times*, 21 June 1998.

———. "U.S. Salmon Plan Could Lead to Removal of Dams." *Los Angeles Times*, 22 December 2000, A3.

Navarro, Peter. "Gas and Pipeline Costs Fueled Electricity Crisis." *Los Angeles Times*, 19 January 2001, B9.

Neuman, W. Russell. *The Paradox of Mass Politics: Knowledge and Opinion in the American Electorate*. Cambridge, Mass.: Harvard University Press, 1986.

"Next, the Oil Recession?" *Newsweek*, 3 December 1973, 87–90.

Nie, Norman H., and James N. Rabjohn. "Revisiting Mass Belief Systems Revisited: Or, Doing Research Is Like Watching a Tennis Match." *American Journal of Political Science* 23 (1979): 139–75.

Nie, Norman H., and Kristi Andersen. "Mass Belief Systems Revisited: Political Change and Attitude Structure." *Journal of Politics* 36 (1974): 541–91.

Nie, Norman H., Sidney Verba, and John R. Petrocik. *The Changing American Voter*. Cambridge, Mass.: Harvard University Press, 1976.

Norris, Pippa. "The Gender Gap: A Cross-National Trend?" In *The Politics of the Gender Gap*, edited by Carol M. Mueller. Newbury Park, Calif.: Sage, 1988.

Nunn, Clyde Z., Harry J. Crockett, Jr., and J. Allen Williams, Jr. *Tolerance for Nonconformity: A National Survey of Americans' Changing Commitment to Civil Liberties*. San Francisco: Jossey-Bass, 1978.

Nye, David E. *Consuming Power: A Social History of American Energies*. Cambridge, Mass.: MIT Press, 1998.

O'Dell, John. "Power Crisis Is a Weapon in Electric Car Debate." *Los Angeles Times*, 19 January 2001, C1.

O'Fallon, John E. "Deficiencies in the Air Quality Act of 1967." In *Air Pollution Control*, edited by Clark C. Havighurst. Dobbs Ferry, N.Y.: Oceana, 1969, 79–100.

"Oil: When Is a Ban Not a Ban?" *Newsweek*, 26 June 1967, 57, 59.

"Oil's High Profits Abroad Bring Trouble in the US." *Oil and Gas Journal*, 6 May 1974, 93–96.

"Opinion Roundup: Nuclear Power." *Public Opinion*, March 1986, 26.

Paddock, Richard C. "Drilling Advance Rekindles Santa Barbara Oil Wars." *Los Angeles Times*, 5 December 1994, A1, A28–29.

Paehlke, Robert C. *Environmentalism and the Future of Progressive Politics*. New Haven, Conn.: Yale University Press, 1989.

Page, Benjamin I., and Robert Y. Shapiro. *The Rational Public: Fifty Years of Trends in Americans' Policy Preferences*. Chicago: University of Chicago Press, 1992.

Petrocik, John R. "Comment: Reconsidering the Reconsiderations of the 1964 Change in Attitude Consistency." *Political Methodology* 5 (1978): 361–68.

Pew Research Center for the People and the Press. "Rising Price of Gas Draws Most Public Interest in 2000." 25 December 2000 (http://www.people-press.org).

Philander, S. George. *Is the Temperature Rising? The Uncertain Science of Global Warming.* Princeton, N.J.: Princeton University Press, 1998.

Phillips, Kevin. *Boiling Point: Republicans, Democrats, and the Decline of Middle-Class Prosperity.* New York: Random House, 1993.

Pierce, John C. "Party Identification and the Changing Role of Ideology in American Politics." *Midwest Journal of Political Science* 14 (1970): 25–42.

———. "The Relationship between Linkage Salience and Linkage Organization in Mass Belief Systems." *Public Opinion Quarterly* 39 (1975): 102–10.

Pine, Art. "Climbing Gas Prices Expected to Be Hot Topic in Campaign." *Los Angeles Times,* 14 March 2000, A16.

"Pollution Alert." *Newsweek,* 5 December 1966, 66–67.

Poole, Keith T., and L. Harmon Ziegler. *Women, Public Opinion, and Politics: The Changing Political Attitudes of American Women.* New York: Longman, 1985.

Portney, Paul R. "Natural Resources and the Environment." In *The Reagan Record,* edited by John L. Palmer and Isabel V. Sawhill. Cambridge, Mass.: Ballinger, 1984.

"Power: Scrounging for Fuel." *Newsweek,* 14 September 1970, 89.

"Power: Sluggish Atom." *Newsweek,* 23 June 1969, 81–82.

President's Commission on the Accident at Three Mile Island. *The Need for Change: The Legacy of TMI.* Washington, D.C.: Government Printing Office, 1979 .

Rankin, William L., Stanley M. Nealey, and Barbara Desow Melber. "Overview of National Attitudes toward Nuclear Energy: A Longitudinal Analysis." In *Public Reaction to Nuclear Power: Are There Critical Masses?* edited by William R. Freudenburg and Eugene A. Rosa. Boulder, Colo.: Westview, 1984, 41–67.

Reese, Stephen D., John A. Daly, and Andrew P. Hardy. "Economic News on Network Television." *Journalism Quarterly* 64 (1987): 137–44.

Reisman, George. "Government's Oil Prices." *Orange County Register,* 2 April 2000, 2.

Riccardi, Nicholas, and Steve Berry. "Deregulation Didn't Foster Competition." *Los Angeles Times,* 7 February 2001, A1.

Richman, Al. "The Polls: Public Attitudes towards the Energy Crisis." *Public Opinion Quarterly* 43 (1979): 576–85.

Risling, Greg. "State's Wholesale Power Suppliers See Surge in Profits." *Santa Barbara News-Press,* 17 December 2000, A6.

Rosa, Eugene A., Marvin E. Olsen, and Don A. Dillman. "Public Views toward National Energy Policy Strategies: Polarization or Compromise?" In *Public Reactions to Nuclear Power: Are There Critical Masses?* edited by William R. Freudenburg and Eugene A. Rosa. Boulder, Colo.: Westview, 1984, 69–93.

Rosa, Eugene A., and Riley E. Dunlap. "Nuclear Power: Three Decades of Public Opinion." *Public Opinion Quarterly* 58 (1994): 295-325.

Rosa, Eugene A., and William R. Freudenburg. "Nuclear Power at the Crossroads." In *Public Reactions to Nuclear Power: Are There Critical Masses?* edited by William R. Freudenburg and Eugene A. Rosa. Boulder, Colo.: Westview, 1984, 18–19.

Rosenbaum, Walter A. *Environmental Politics and Policy.* 2d ed. Washington, D.C.: Congressional Quarterly, 1991.

Rosenblatt, Robert A. "Energy Crisis: Oil Firms, U.S. Caused Shortage." *Los Angeles Times*, 17 July 1973.

Rosenblatt, Robert A., and Doyle McManus. "Clinton Taps Oil Reserves to Ease Shortage." *Los Angeles Times*, 23 September 2000, A1.

Saad, Lydia. "Americans Mostly 'Green' in the Energy vs. Environment Debate." Gallup Poll Release, 16 March 2001.

Scammon, Richard M., and Ben J. Wattenberg. *The Real Majority*. New York: Coward, McCann and Geoghegan, 1970.

Schrag, Peter. "Blackout: Did California Just Mess It Up Badly, or Is Deregulation in Electricity an Inherently Bad Idea?" *The American Prospect*, 26 February 2001, 29–33.

Schuman, Howard, and Stanley Presser. *Questions and Answers in Attitude Surveys: Experiments on Question Form, Wording, and Context*. New York: Academic, 1981.

"Shadow War on the Economic Front." *Newsweek*, 19 June 1967, 68–70.

Shafer, Byron E., and William J. M. Claggett. *The Two Majorities: The Issue Context of Modern American Politics*. Baltimore: Johns Hopkins University Press, 1995.

Shapiro, Robert Y., and Harpreet Mahajan. "Gender Differences in Policy Preferences: A Summary of Trends from the 1960s to the 1980s." *Public Opinion Quarterly* 49 (1986): 42–61.

Shogren, Elizabeth. "Wind Farm Called Threat to Condors." *Los Angeles Times*, 14 September 1999, A3.

———. "Bush Drops Pledge to Curb Emissions." *Los Angeles Times*, 14 March 2001, A1.

Simmons, Wendy W. "Despite Dire Predictions of Global Warming, Americans Have Other Priorities." Gallup Poll Release, 20 February 2001.

Slater, Eric. "In Race, Rising Gas Prices Are Make-or-Brake Issue." *Los Angeles Times*, 24 June 2000, A1.

Smith, Eric R. A. N. *How Political Activists See Offshore Oil Development: An In-Depth Investigation of Attitudes on Oil Development*. Camarillo, Calif.: U.S. Minerals Management Service, 1998.

———. *The Unchanging American Voter*. Berkeley: University of California Press, 1989.

———. "What Is Public Opinion?" *Critical Review* 10 (1996): 95–105.

Smith, Eric R. A. N., and Sonia R. Garcia. "Californians' Attitudes toward Energy Policy." Paper delivered at the 1994 Annual Meeting of the Western Political Science Association, Albuquerque, New Mexico, March 1994.

———. "Evolving California Opinion on Offshore Oil Development." *Ocean and Coastal Management* 26 (1995): 41–56.

Smith, Tom W. "The Polls: Gender and Attitudes toward Violence." *Public Opinion Quarterly* 48 (1984): 384–96.

Sniderman, Paul M., with Michael Gray Hagen. *Race and Inequality: A Study in American Values*. Chatham, N.J.: Chatham House, 1985.

Sollen, Robert. *An Ocean of Oil: A Century of Political Struggle over Petroleum off the California Coast*. Juneau, Ala.: Denali, 1998.

Solomon, Lawrence S., Donald Tomaskovic-Devey, and Barbara J. Risman. "The Gender Gap and Nuclear Power: Attitudes in a Politicized Environment." *Sex Roles* 21 (1989): 401–14.

Stamm, Keith R., Fiona Clark, and Paula R. Eblacas. "Mass Communication and Public Understanding of Environmental Problems: The Case of Global Warming." *Public Understanding of Science* 9 (2000): 219–37.

Stanley, Harold W., and Richard G. Niemi. *Vital Statistics for American Politics.* 3d ed. Washington, D.C.: Congressional Quarterly, 1992.

———. *Vital Statistics on American Politics.* 5th ed. Washington, D.C.: Congressional Quarterly, 1995.

Steinhart, Carol E., and John S. Steinhart. *Blowout: A Case Study of the Santa Barbara Oil Spill.* Belmont, Calif.: Duxbury, 1972.

Stimson, James A. *Public Opinion in America.* Boulder, Colo.: Westview, 1991.

Stone, Gerald C., and Elinor Grusin. "Network TV as a Bad News Bearer." *Journalism Quarterly* 61 (1984): 517–23.

Stone, Gerald C., Barbara Hartung, and Dwight Jensen. "Local TV News and the Good-Bad Dyad." *Journalism Quarterly* 64 (1987): 37–44.

Stouffer, Samuel A. *Communism, Conformity, and Civil Liberties: A Cross-Section of the Nation Speaks Its Mind.* Garden City, N.Y.: Doubleday, 1955.

Strand, Paul J. "The Energy Issue: Partisan Characteristics." *Environment and Behavior* 13 (1981): 509–19.

Sullivan, John L., James E. Piereson, and George E. Marcus. "Ideological Constraint in the Mass Public: A Methodological Critique and Some New Findings." *American Journal of Political Science* 22 (1978): 233–49.

Tamaki, Julie. "Groups Look for Silver Lining in Energy Crisis." *Los Angeles Times,* 18 March 2001, A38.

Tashchian, Roobina Ohanian, and Mark E. Slama. "Survey Data on Attitudes and Behaviors Relevant to Energy: Implications for Policy." In *Families and the Energy Transition,* edited by John Byrne, David A. Schulz, and Marvin B. Sussman. New York: Haworth, 1985, 29–51.

Taylor, Dorceta E. "Blacks and the Environment: Toward an Explanation of the Concern Gap between Blacks and Whites." *Environment and Behavior* 21 (1989): 175–205.

"Terrible Swift Sword." *Newsweek,* 19 June 1967, 24–34.

"There Goes the Power." *Newsweek,* 10 August 1970, 65.

Thompson, P. T., and J. MacTavish. *Energy Problems: Public Beliefs, Attitudes and Behaviors.* Allendale, Mich.: Grand Valley State College, Urban and Environmental Studies Institute, 1976.

Thurow, Lester. *The Zero-Sum Economy: Distribution and the Possibilities for Economic Change.* New York: Penguin, 1980.

Tonelson, Alan, and Beth A. Lizut. "If We Kicked the Oil Habit, Saddam Wouldn't Menace Us." *Washington Post,* 15 September 1996.

"Top Oil Firms' 1973 Profits Jump 52.7%." *Oil and Gas Journal* 72 (18 February 1974): 32–34.

Tremblay, Kenneth R., Jr., and Riley E. Dunlap. "Rural-Urban Residence and Concern with Environmental Quality: A Replication and Extension." *Rural Sociology* 43 (1978): 474–91.

Trumbo, Craig. "Longitudinal Modeling of Public Issues: An Application of Agenda-Setting Processes to the Issue of Global Warming." *Journalism and Mass Communication Monographs* 152 (August 1995).

Tyndall, J. "On Radiation through the Earth's Atmosphere." *Philosophical Magazine* 4 (1983): 200.

"Up the Wall." *Newsweek,* 18 March 1974, 99–100.

U.S. Census Bureau. *1990 Census of Population. General Population Characteristics, California.* Washington, D.C.: Government Printing Office, 1992.

————. *Statistical Abstract of the United States, 1998.* 118th ed. Washington, D.C.: Government Printing Office, 1998.

U.S. Energy Information Administration, *Annual Energy Review 1997.* Washington, D.C.: Government Printing Office, 1998.

van der Pligt, Joop, J. Richard Eiser, and Russell Spears. "Attitudes toward Nuclear Energy: Familiarity and Salience." *Environment and Behavior* 18 (January 1986): 75–93.

————. "Nuclear Waste: Facts, Fears, and Attitudes." *Journal of Applied Social Psychology* 17, no. 5 (1987): 453–70.

Van Liere, Kent D., and Riley E. Dunlap. "The Social Bases of Environmental Concern: A Review of Hypotheses, Explanations and Empirical Evidence." *Public Opinion Quarterly* 44 (1980): 181–97.

Vartabedian, Ralph. "Dipping into the Strategic Petroleum Reserve." *Los Angeles Times,* 9 May 1996, D1.

Verba, Sidney, and Norman H. Nie. *Participation in America.* New York: Harper and Row, 1972.

Vogel, Nancy. "Crisis Making Leaders Rethink End of Controls." *Los Angeles Times,* 4 August 2000, A1.

Vogel, Nancy, and Dan Morain. "Governor, Legislators Moving toward Bailout of Utilities." *Los Angeles Times,* 5 January 2001, A1.

Warner, Keith D. "Extinction Is Forever, Dams Are Not." *Los Angeles Times,* 30 June 2000, B8.

"Warning: Low Voltage." *Newsweek,* 18 May 1970, 123.

Webber, David J. "Is Nuclear Power Just Another Environmental Issue? An Analysis of California Voters." *Environment and Behavior* 14 (1982): 72–83.

Wellock, Thomas Raymond. *Critical Masses: Opposition to Nuclear Power in California, 1958–1978.* Madison: University of Wisconsin Press, 1998.

Welsh, Nick. "All's Well That Ends Wells?" *The Independent,* 14 March 1996, 23–25.

Whitman, David. "The Coal Hard Facts." *U.S. News & World Report,* 26 March 2001, 17–18.

Wicker, Tom. *One of Us: Richard Nixon and the American Dream.* New York: Random House, 1991.

Wildavsky, Aaron. *The Rise of Radical Egalitarianism.* Washington, D.C.: American University Press, 1991.

————. "Risk Perception." *Risk Analysis* 11 (1991): 15.

Wildavsky, Aaron, and Karl Dake. "Theories of Risk Perception: Who Fears What and Why." *Daedalus* 41 (1990): 41–60.

Wilder, Robert Jay. *Listening to the Sea: The Politics of Improving Environmental Protection.* Pittsburgh, Pa.: University of Pittsburgh Press, 1998.

Williams, James C. *Energy and the Making of Modern California.* Akron, Ohio: University of Akron Press, 1997.

Wilson, Thomas C. "Cohort and Prejudice: Whites' Attitudes toward Blacks, Hispanics, Jews, and Asians." *Public Opinion Quarterly* 60 (Summer 1996): 253–74.

Wolverton, David, and Donald Vance. "Newspaper Coverage of Proposals for Rate Increases by Electric Utility." *Journalism Quarterly* 64 (1986): 581–84.

Woolley, John T. "Using Media Reports as Indicators of Policy Processes." *American Journal of Political Science* 44 (2000): 156–73.

Yergin, Daniel. *The Prize: The Epic Quest for Oil, Money and Power*. New York: Touchstone, 1991.

"You Say Gas Prices Are Up? Let the Political Games Begin." *Los Angeles Times*, 1 May 1996, B8.

Zaller, John R. "Bringing Converse Back In: Modeling Information Flow in Political Campaigns." *Political Analysis* 1 (1989): 181–234.

———. "Information, Values, and Opinion." *American Political Science Review* 85 (1991): 1215–38.

———. *The Nature and Origins of Mass Opinion*. Cambridge: Cambridge University Press, 1992.

Index

About the Author

Eric R. A. N. Smith is associate professor of political science at the University of California, Santa Barbara, and is an affiliated faculty member of the Environmental Studies Program at UCSB. He is the author of *The Unchanging American Voter* (1989) and coauthor of *Dynamics of Democracy*, 3d ed. (2001).